Lecture Notes in Control and Information Sciences 386

Editors: M. Thoma, F. Allgöwer, M. Morari

Dan Huang, Sing Kiong Nguang

Robust Control for Uncertain Networked Control Systems with Random Delays

Authors

Dr. Dan Huang

University of Auckland
Department of Electrical & Computer Engineering
38 Princes Street
Auckland
New Zealand
E-mail: dan.huang@auckland.ac.nz

Prof. Sing Kiong Nguang

University of Auckland
Department of Electrical & Computer Engineering
38 Princes Street
Auckland
New Zealand
E-mail: sk.nguang@auckland.ac.nz

ISBN 978-1-84882-677-9 e-ISBN 978-1-84882-678-6

DOI 10.1007/978-1-84882-678-6

Lecture Notes in Control and Information Sciences ISSN 0170-8643

Library of Congress Control Number: Applied for

Typeset & Cover Design: Scientific Publishing Services Pvt. Ltd., Chennai, India.

Printed in acid-free paper

5 4 3 2 1 0

springer.com

Preface

Networked control systems (NCSs) are a type of distributed control systems where sensors, actuators, and controllers are interconnected through a communication network. This system setup has the advantage of low cost, flexibility, and less wiring, but it also inevitably invites some delays and data loss into the design procedure.

The focus of this book is to address the problem of analysis and design of NCSs when the communication delays are varying in a random fashion. This random feature of the time delays is typical for commercially used networks, such as a DeviceNet^{TM}(which is a controller area network (CAN)) and EtherNet^{TM}network. Models for communication network delays are first developed, in which Markov processes are used to model these random network-induced delays. Based on such models, we establish novel methodologies for stability analysis, control with disturbance attenuation, filter design, and fault estimation for a class of uncertain linear/nonlinear uncertain NCSs with random communication network-induced delays in both sensor-to-controller and controller-to-actuator channels. Data packet dropouts in the communication channels also have been taken into consideration in the modeling and design procedure.

The main technique used in this book is based on the Lyapunov–Razumikhin method, which results in delay-dependent controllers. We first consider the design problems for uncertain linear NCSs. In this case, state feedback controllers and dynamic output feedback controllers are designed to satisfy both stability and disturbance attenuation requirements for this class of NCSs. A robust filter design problem is also considered. Moreover, a robust fault estimator that ensures the fault estimation error is less than a prescribed performance level is designed. We further go on to address the control problems for uncertain nonlinear NCSs. The nonlinear plant is first described by the Takagi–Sugeno (T-S) fuzzy model. Based on this model, stability analysis, disturbance attenuation, filter design, and fault estimation problems are studied for uncertain nonlinear NCSs in parallel. It should be noted that system uncertainties, disturbances and noises are addressed in both cases.

The existence of such controllers, filters, and fault estimators are given in terms of the solvability of bilinear matrix inequalities (BMIs). Iterative algorithms are proposed to change this non-convex problem into quasi-convex optimization problems, which can be solved effectively by available mathematical tools.

Finally, to demonstrate the effectiveness and advantages of the proposed design methodologies in this book, numerical examples are given in each designed control systems. The simulation results show that the proposed design methodologies can achieve the prescribed performance requirements.

Auckland, New Zealand
April 2009

Dan Huang
Sing Kiong Nguang

Contents

Abbreviations XI

1 Introduction 1
1.1 Introduction of Networked Control Systems 1
1.2 Fundamental Issues with Networked Control Systems 3
1.3 Recent Works on Networked Control Systems 6
1.4 Research Motivation 11
1.5 Contribution of the Book 13
1.6 Book Outline 14

2 Modeling of Networked Control Systems 17
2.1 Formulation of Networked Control Systems 17
2.2 Modeling of Random Network-induced Delays 19

Part I: Linear Uncertain Networked Control Systems

3 State Feedback Control of Uncertain Networked Control Systems 25
3.1 Introduction 25
3.2 Problem Formulation and Preliminaries 26
3.3 Main Result 27
3.4 Numerical Example 32
3.5 Conclusion 33

4 Dynamic Output Feedback Control for Uncertain Networked Control Systems 37
4.1 System Description and Problem Formulation 37
4.2 Main Result 39
4.3 Numerical Example 49
4.4 Conclusion 50

5 Robust Disturbance Attenuation for Uncertain Networked Control Systems ... 53
5.1 Introduction ... 53
5.2 System Description and Problem Formulation ... 54
5.3 Main Result ... 57
5.4 Numerical Example ... 60
5.5 Conclusion ... 63

6 Robust Filter Design for Uncertain Networked Control Systems ... 65
6.1 Introduction ... 65
6.2 System Description and Problem Formulation ... 65
6.3 Main Result ... 68
6.4 Numerical Example ... 70
6.5 Conclusion ... 72

7 Robust Fault Estimator Design for Uncertain Networked Control Systems ... 73
7.1 Problem Formulation and Preliminaries ... 74
7.2 Main Result ... 76
7.3 Numerical Example ... 80
7.4 Conclusion ... 82

Part II: Nonlinear Uncertain Networked Control Systems

8 Takagi–Sugeno Fuzzy Control System ... 87
8.1 Takagi-Sugeno Fuzzy Modeling ... 88
8.2 Takagi-Sugeno Fuzzy Controller ... 90

9 State Feedback Control for Uncertain Nonlinear Networked Control Systems ... 93
9.1 Introduction ... 93
9.2 Problem Formulation and Preliminaries ... 93
9.3 Main Result ... 96
9.4 Numerical Example ... 102
9.5 Conclusion ... 104

10 Dynamic Output Feedback Control for Uncertain Nonlinear Networked Control Systems ... 107
10.1 Problem Formulation and Preliminaries ... 107
10.2 Main Result ... 110
10.3 Numerical Example ... 113
10.4 Conclusion ... 115

11 Robust Disturbance Attenuation for Uncertain Nonlinear Networked Control Systems ... 117
11.1 Problem Formulation and Preliminaries ... 117
11.2 Main Result ... 120
11.3 Numerical Example ... 123
11.4 Conclusion ... 126

12 Robust Fuzzy Filter Design for Uncertain Nonlinear Networked Control Systems ... 129
12.1 Problem Formulation and Preliminaries ... 129
12.2 Main Result ... 132
12.3 Example ... 135
12.4 Conclusion ... 136

13 Fault Estimation for Uncertain Nonlinear Networked Control Systems ... 137
13.1 Problem Formulation and Preliminaries ... 137
13.2 Main Result ... 140
13.3 Numerical Example ... 142
13.4 Conclusion ... 144

14 Conclusions ... 145
14.1 Summary of the Book ... 145
14.2 Future Research Work ... 146

A Mathematical and Background Knowledge ... 147
A.1 Linear Matrix Inequality ... 147
A.1.1 LMI Problems ... 148
A.1.2 The Schur Complement ... 148
A.2 Continuous-time Markov Process ... 149
A.3 Lemmas ... 150

References ... 151

Index ... 159

Abbreviations

NCS	Networked Control System
CAN	Controller Area Network
QoS	Quality-of-Service
ZOH	Zero Order Hold
LQG	Linear Quadratic Gaussian
MAC	Medium Access Control
MATT	Maximum Allowable Transfer Time
FLC	Fuzzy Logic Control
T-S	Takagi-Sugeno
LMI	Linear Matrix Inequality
BMI	Bilinear Matrix Inequality
MJLS	Markovian Jump Linear System
LQR	Linear Quadratic Regulator
FDI	Fault Detection and Isolation
ILMI	Iterative Linear Matrix Inequality
GEVP	Generalized EigenValue Problem

Chapter 1
Introduction

1.1 Introduction of Networked Control Systems

The point-to-point architecture is the traditional communication architecture for control systems, that is, sensors and/or actuators are connected to controllers via wires. In recent years, due to the expansion of physical setups and functionality, a traditional point-to-point architecture is no longer able to meet new requirements, such as modularity, integrated diagnostics, quick and easy maintenance, and low cost. Such requirements are particularly demanding in the control of complex control systems [1, 2, 12] and remote control systems [3, 4, 10, 13].

To satisfy these new requirements, common-bus network architectures have been introduced. The common-bus network architectures can improve the efficiency, flexibility and reliability of integrated applications, and reduce installation, reconfiguration and maintenance time and costs. In recent years, therefore, it gives rise to the so-called network-based control systems, or networked control systems (NCSs) [5, 6, 7, 8, 41].

Figure 1.1 shows an application of NCSs in medical/health-care systems. In this system, each patient is monitored by certain devices connected to the patient. These devices send out the patient's medical information to a central host machine which communicates through networks or satellites with a remote database interconnected with a tuning device. Once the medical information is compared with that stored in the database, indicating any abnormal symptoms, the tuning device then sends out regulating signals which in turn stimulate those devices that are connected to each patient. This system provides a sound solution to the widely existing shortage of health-care staff.

In general, NCSs are a type of distributed control systems where sensors, actuators, and controllers are interconnected through a communication network as shown in Figure 1.2. Sensors measure states of the plant and transmit these states over the communication network to controllers. The controllers receive these states, and calculate appropriate control actions and send them to actuators over the communication network. Actuators receive control actions and control the plant appropriately. Due to its low cost, flexibility, and less wiring, NCSs are rapidly increasing

D. Huang and S.K. Nguang: Robust Ctrl. for Uncertain Networked Ctrl. Sys., LNCIS 386, pp. 1–15.
springerlink.com

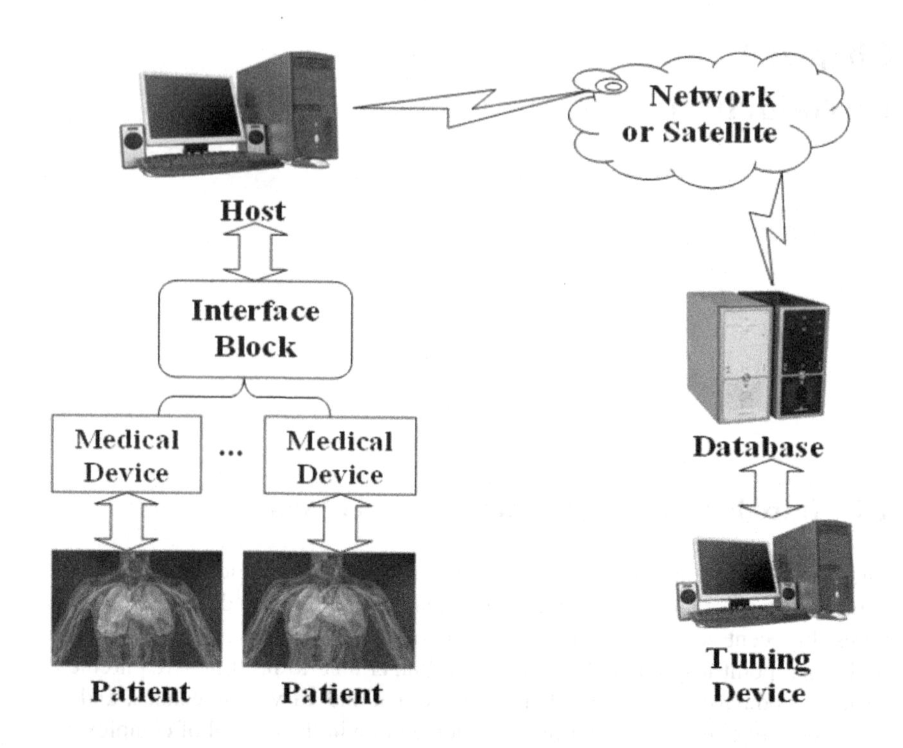

Fig. 1.1 A medical application of NCSs

in industrial applications, including telecommunications, remote process control, altitude control of airplanes and so on [3, 4, 2, 9, 10, 11, 12, 13].

It can be seen from the block diagram in Figure 1.2 that in NCSs the closed-loops are closed via communication networks. The insertion of the communication network in the feedback control loop makes the analysis and design of systems more complex than the traditional point-to-point architecture. The network can introduce unreliable and time-dependent levels of service in terms of, for example, delays, jitter, or losses. Quality-of-Service (QoS) [14] can ameliorate the real-time network behavior, but the network behavior is still subject to interference (especially in wireless media), routing transients, to aggressive flows. It is also noteworthy that protocols providing QoS are not prevalent in all industrial networks. In general, network vagaries can jeopardize the stability, safety, and performance of the units in a physical environment.

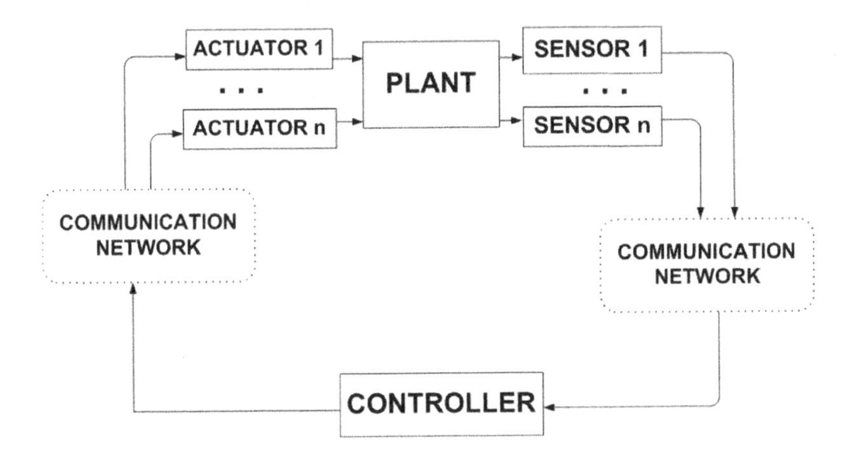

Fig. 1.2 Typical setup of NCSs

1.2 Fundamental Issues with Networked Control Systems

The following two issues are the most challenging problems with NCSs that need to be properly addressed to ensure the stability and performance of the closed-loop systems.

1. The first issue is the network-induced delay, including sensor-to-controller delay and controller-to-actuator delay, that happens when data exchange happens among devices connected by the communication network, which will deteriorate the system performance as well as stability. This delay, depending on the network characteristics such as network load, topologies, routing schemes, etc., can be constant, time-varying, or even random.
2. The second issue is the data packet dropouts. In the NCS, data is sent through the network in packets. Due to this network characteristic, therefore, any continuous-time signal from the plant is first sampled to be carried over the communication network. Chances are that those packets can be lost during transmission because of uncertainty and noise in communication channels. It may also occur at the destination when out of order delivery takes place.

It is straightforward to learn that the severity of the network-induced delays is aggravated when data packet dropouts occur during a network transmission. Furthermore, in the case of multiple-packet transmission, chances are that part/none of the packets could arrive by the time of control calculation, which makes the analysis and control of NCS more difficult, if not impossible.

We now consider a generic schematic diagram of NCS as shown in Figure 1.3. In this system, the continuous output signal $x(t)$ is sampled by an ideal sensor/sampler h^s at a sampling rate of $1/T_S$. Then digital signal is transmitted through a communication network after experiencing network-induced delay τ_{sc}^{k-1} and a zero order hold (ZOH) to the controller. The clock-driven controller generates a digital control signal $\hat{u}_k$ at the rate of $1/T_S$ and again transmits via the network, after certain delay τ_{ca}^k, back to a ZOH of the actuator side which is event-driven.

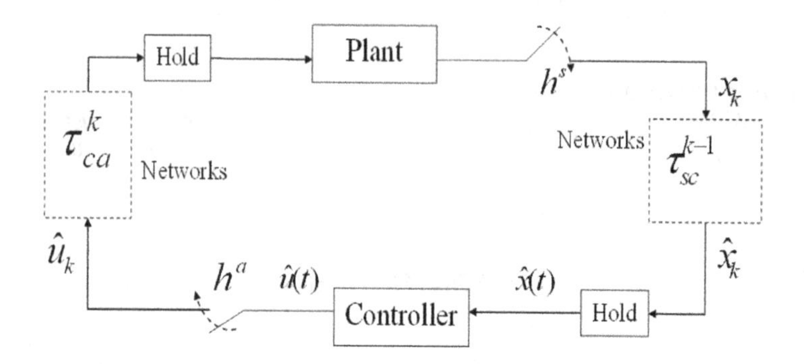

Fig. 1.3 Schematic diagram of a networked control system

The timing mechanism diagram of this kind of configuration of NCS is drawn in Figure 1.4. We assume that each data can reach its destination within one sampling period $T_S = t_k - t_{k-1}$, that is, network-induced delay τ_{sc}^{k-1} and τ_{ca}^k is less than T_S.

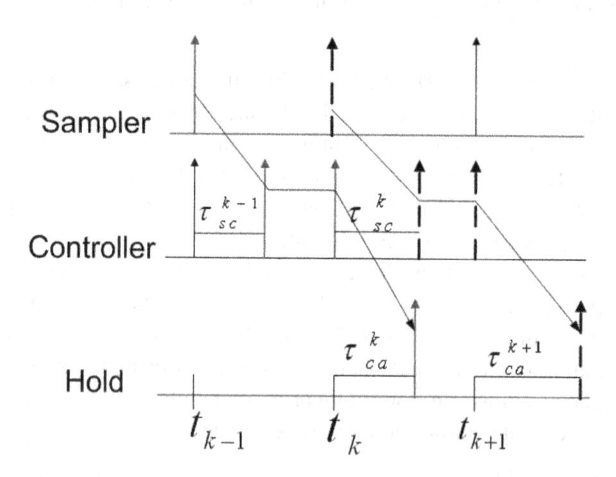

Fig. 1.4 Timing mechanism diagram of a networked control system

The operation of the NCS runs as follows. At the previous time instant t_{k-1}, the output signal $x(t)$ is measured by a clock-driven sensor. (which is in solid line.) The sampled signal reaches the ZOH at the controller side at time $t_{k-1} + \tau_{sc}^{k-1}$. The controller is also clock-driven and at time t_k it generates a control signal $\hat{u}_k$ using the most recently arrived signal. This control signal is received by the ZOH at the actuator side after τ_{ca}^{k} and will be held until the new control signal (green one) arrives at the instant $t_{k+1} + \tau_{ca}^{k+1}$. This means a control signal is held for the period $(t_k + \tau_{ca}^{k}, t_{k+1} + \tau_{ca}^{k+1}]$. We can also learn from the system formulation that the system contains both continuous-time and discrete-time signals, where a hybrid systematic synthesis approach is needed.

In a real network, however, the network-induced delay is much more volatile than mentioned in the previous section which is under ideal assumption. Actually, sensor-to-controller delay τ_{sc}^{k} and controller-to-actuator delay τ_{ca}^{k} vary greatly because of the medium access control (MAC) protocol of the network, such as its topologies, routing schemes, etc. In the following section, two types of networks in terms of different MAC protocols will be introduced.

1. **Cyclic service (scheduling) network.** Token passing protocol, which appears in token bus (e.g. IEEE standard 802.4, SAE token bus, PROFIBUS™), token ring (e.g. IEEE standard 802.5 [15]), and the fiber distributed data interface MAC architectures [16], and time division multiple access protocol which is used in FireWire® (IEEE 1394)[17], are two of the most commonly employed protocols in scheduling network. In such network, data is transmitted in a cyclic order and the variation of τ_{sc}^{k} and τ_{ca}^{k} can be periodic and deterministic under error-free communication assumption [18]. This periodic property however can be destroyed by the following detrimental factors.

   a. The first factor that destroys the periodic property is caused by communication errors. When communication errors happen, a data packet may not be able to be transmitted to a destination component after the next sampling period. This means that there is no updated measurement for generating next control action. This phenomenon is called vacant sampling [8]. On the other hand, two data messages may arrive at the destination during the same sampling time period and because of the nature of the component, only the most recent data packet is used and the previous one is then rejected. This situation is referred to as message rejection or data packet dropouts [8].

      Figure 1.5 illustrates the timing mechanism that incorporates the aforementioned network-induced effects, i.e., vacant sampling and data dropout in network transmission. The output signal measured at time instant t_k (marked in broken line) is lost due to these two network communication errors.

      From Figure 1.5, it is also noteworthy that the data packet dropouts can be viewed as a prolonged network-induced delays that are at least longer than one sampling period. This would be an interesting feature in the modeling of the time-delays as data packet dropouts.

   b. The other factor that destroys the periodic property is caused by clock drift (network jitter). Even though transmissions on scheduling networks should

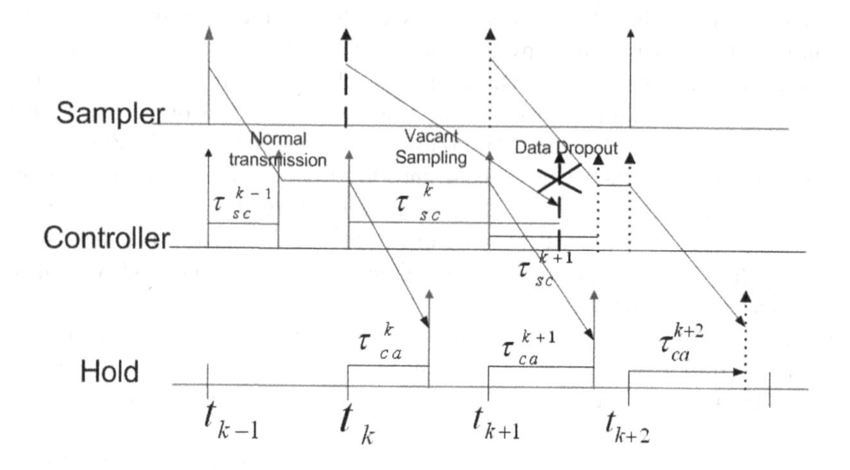

Fig. 1.5 Timing mechanism diagram of vacant sampling and data dropout in network transmission

be periodic in theory, in practice they are up to the network jitter (defined as the time-variation in actual start time of actions) which causes the difference between the sampling period of different components due to differences in the material that generates clock signals. Clock drift changes the probabilities of the occurrences of vacant sampling and message rejection [19]. System components can perform clock synchronization to reduce the clock drift problem [20]. It should also be noted that in Figure 1.4, the sensor and controller are synchronized with the same sampling period of T_S. This is a common practice in industrial control systems, for example, field-bus control systems. Clock synchronization which is also discussed in [21] is an essential function to make the control system as it should be.

2. **Random access network.** Carrier sense multiple access protocol is most often used in random access network whose application includes DeviceNet™(which is a CAN)[22]), EtherNet™[15], and Internet [3]. In random access network, τ_{sc}^k and τ_{ca}^k are stochastic processes [23]where stochastic approaches are needed to model the behaviors of τ_{ca}^k and τ_{sc}^k.

1.3 Recent Works on Networked Control Systems

Our focus in this research work is on the control *through* networks, not control *of* networks. Therefore, our design objectives involve the performance and stability of a target NCS rather than of the network. In this regard, we present the following recent development of NCSs on this issue.

In the research of NCSs, stability analysis is one of the most concerned areas, therefore much effort has been devoted to this problem; for examples, see [24, 25, 26, 27, 28, 6, 41, 46, 47, 50, 55, 34, 35, 36, 37, 38, 39]. Numerous techniques have been developed to address the stability problems, which will be introduced in the following text. Emphasis is mainly on the modeling of network-induced delays.

Based on the constructed models, state feedback and output feedback controllers are designed to meet the stability requirements. In addition, maximum allowable transfer time (MATT) to guarantee system stability is also discussed in [55, 6, 29].

Some efforts have also been made in the realm of performance control for NCSs. $\mathscr{H}_\infty$ disturbance attenuation analysis for a class of NCSs was presented in [46]. Robust filtering problem was discussed in [30, 32, 33]. In [64], guaranteed cost control for NCSs was considered. The authors also dealt with networked systems under noise disturbance in [31]. Compared with the research work in the stability analysis and control, performance control for NCSs lags behind.

The study of NCSs is an interdisciplinary research area, combining both network and control theories. Hence, in order to guarantee the stability and performance of NCSs, analysis and design tools based on both network and control parameters are required.

It is worth mentioning that the research on NCSs cannot simply treated as time-delay systems due to data packet dropout problems and other network features in the NCSs. Various methods have been applied, based on different types of network configurations, to treat the delay and data packet dropout problems. In these methods, some assumptions have been made to derive generic control techniques for NCSs. To name a few:

- network transmissions are error-free,
- both sensor-to-controller delay and controller-to-actuator delay are constant,
- time-delay is less than one sampling period,
- network traffic cannot be overloaded.

Based on such assumptions, different control techniques have been developed for the control of NCSs, and some of which are described as follows:

1. **Sampled-data system/hybrid approach.** NCS is basically a hybrid system in nature, which involves continuous plant and event-driven or time-driven devices (digital controller, sampler, and holder). Data is sent through the network as atomic units called packets. Therefore any continuous-time signal must be appropriately sampled to be carried over a network. Hence there are some similarities between NCSs and sampled-data systems due to the sampling effect. Sampled-data system formulation of NCS is quite general [40, 6, 41, 42, 43] and can also capture this hybrid characteristic and many other network-induced features, including time-delay, packet dropout, and so on. It also provides a natural way to describe the so-called "communication sequence" [45].

   An analysis of NCS is considered under hybrid framework with one channel time-delay in [41] and [6]. In [8, 19], the authors prove that two time-delays cannot be lumped together for the cases where vacant sampling and message rejection happens. In [44], the authors propose a modeling problem of NCSs for multi-variable linear systems with distributed asynchronous sampling. In [42], the authors consider a robust control method by transforming the sampled-data system into a continuous-time system with control delay. However this transformation is under the assumption that the sampling period is infinitly small. In [40], a less conservative time-delay dependent stability result is obtained by transforming time-delay in the discrete-time subsystem into its continuous-time subsystems of the sampled-data systems using a new Lyapunov function. It also removes the limitation that on the sampling period in [40] and the assumption used in [21] that the time-delay from sensor to actuator is less than one sampling period.

2. **Switched system approach.** A switched system means a hybrid dynamical system consisting of a finite number of subsystems described by differential or difference equations and a logical rule that orchestrates switching between these subsystems. In [46] stability and disturbance attenuation issues for a class of NCSs under uncertain access delay and packet dropout effects are considered in the framework of switched systems. The basic idea is to formulate such NCS as a discrete-time switched system. Then the NCSs' stability and performance problems can be reduced to corresponding problems for the switched systems, which have been studied for decades and for which a number of results are available in the literature [47, 48, 49]. The stability problem of NCSs is also studied in [6] by applying the multiple Lyapunov function method [49].

   The strength of this approach comes from the solid theoretical results existing in the literature for stability, robust performance etc., for switched systems. Nevertheless, it requires the controller works at a higher frequency than the sampling frequency. Furthermore, only sensor-to-controller channel time-delay is accommodated in this work.

3. **Sampling time scheduling approach.** This method is applied to cyclic service networks in which all system component connections on the network are known in advance. In [18], this method is applied by appropriately selecting a long enough sampling period for a discrete-time network-based system so that network delays do not affect the control performance and the systems can remain stable. The control system comprises a time-driven sensor and controller and an event-driven actuator/holder. Furthermore, assumptions are made that the time-delay is less than one sampling period. The control method can only to applied to a one-dimensional system. In [50, 51] this technology has been expanded to multi-dimensional cases.

4. **Augmented deterministic discrete-time model approach.** In [19, 8], an augmented deterministic discrete-time model is proposed for NCSs with periodic network delays. The sensor and controller are time-driven while the actuator is event-driven. The state-space equations of the linear systems are first converted into finite dimensional discrete-time equations to include past values of plant input and output (i.e., delayed variables) as additional states under the assumption that the system input is piecewise constant during each sampling period. Then such equations are combined and rearranged into an augmented state-space equation. The system can be proved as asymptotically stable if all eigenvalues of the

product of the certain augmented state transition matrix of the augmented systems are contained within the unit circle. In [52] this method has been advanced to deal with non-identical sampling periods of sensor and controller.

However, the complexity of the system increases significantly and proportionally to the dimension of the states and inputs as a result of extensive state augmentation, which introduces unrealistic computational time for systems with high orders.

5. **Optimal stochastic control approach.** In [20, 53], an optimal stochastic control approached is presented for the control of a system over a random delay network. The effects of network delays are treated as a linear quadratic gaussian (LQG) problem. The controller and actuator used in this approach is event-driven while the sensor is time-driven. It also requires that the sum of sensor-to-controller and controller-to-actuator time-delay is less than the sampling period, i.e., $\tau_{sc}^{k}+\tau_{ca}^{k} < T_S$, and the information of all the past delays are available. In this approach, two stochastic processes were incorporated into the system state-space equations and the goal is to minimize a cost function of the plant states and inputs. The stability of the network-based system for both independent delays and delays modeled by a Markov chain is discussed using stochastic stability analysis.

   This approach is more realistic to the nature of the time-delays and has shown better performance. However, finding the Markov relation of a delay is a challenging task. Moreover, a large requirement of controller memory to store past information is necessary in this approach and the assumption used in this approach may not be effective for a system with fast response time.

6. **Perturbation approach.** The fundamental idea of the perturbation method [54, 55] is to formulate delay effects as a perturbation of a continuous-time system under the assumption that there is no observation noise. This requires a very small sampling period so that a network-based system can be approximated by a continuous-time model.

   In this method, the system setup includes a time-driven sensor, an event-driven controller and an event-driven actuator. In the feedback loop, however, the network is only used in the sensor to controller channel. This method can be applied to both cyclic service and random access types of networks. However, these networks are restricted to the priority-based networks. The priorities are managed by priority scheduling algorithms [56] which can be either fixed or changeable.

   The advantage of the perturbation method is that it can be applied to a nonlinear system. However, it is not very practical as controller-to-actuator delays cannot be included in this approach.

7. **Queuing approach (predictor-based compensation approach).** Queuing mechanisms are developed which utilize deterministic [57, 58] or probabilistic [1] information of NCSs for control purposes. In the deterministic case, an observer and a predictor are used for the delay compensation of NCSs with random delays based on the past output measurements which are stored in a first-in-first-out queue. This approach has a high demand of plant preciseness and may result in unnecessarily long delays because of queues. In the probabilistic case, the knowledge of the data lengths in a queue is used to improve the prediction.

However, unnecessary delays from queues still remains in both cases. The latter approach also neglects the effect of controller-to-actuator delays.

8. **Moving horizon approach.** In [59], the authors present a control system design strategy for multivariable plants where the controller, sensors and actuators are connected via a digital, data-rate limited, communications channel. In order to minimize bandwidth utilization, a communication constraint is imposed which restricts all transmitted data to belong to a finite set and only permits one plant to be addressed at a time. The authors then emphasized implementation issues and employed moving horizon techniques [60, 61] to deal with both control and measurement quantization issues which mainly occurred in the up-link (from sensors to controller) side where signal quantization is employed to minimize data rate requirements.

   Moving horizon ideas allow one to trade off computational complexity with performance and is particularly suited to protocols where the message size can be manipulated. However, bandwidth is also conserved due to the dynamic optimization of the system with respect to supplying control increments only when they are required.

9. **Fuzzy logic control approach.** In recent years, there has been a rapidly growing popularity of the applications of T-S fuzzy model [62]. The main feature of a T-S fuzzy model is to express the local dynamics of each fuzzy rule by a simple linear-system model. The overall fuzzy model of a system is achieved by fuzzy "blending of the local models with membership functions. Compared with existing NCS modeling methods, this approach does not require the knowledge of exact values of network-induced delays. Instead, it addresses situations involving all possible network-induced delays. Some attempts have also been made in the design of NCSs by applying this fuzzy modeling technology [63, 64, 65].

   In [64], NCSs with random but bounded delays and packets dropout are first modeled as discrete-time jump fuzzy systems. A guaranteed cost state-feedback controller is then designed associated with a piecewise quadratic Lyapunov function. The controller-to-actuator channel delay however is not addressed in this work. In [63], a parity-equation approach and a fuzzy-observer-based approach for fault detection of an NCS are developed based on a T-S fuzzy model that generates the Markov transfer matrix for the NCS. An NCS for servo motor control is implemented on a Profibus-DP network in [65]. The NCS consists of several independent, but interacting, processes running on two separate stations. A remote fuzzy logic controller is proposed to compensate the network-induced delay for a single-input-single-output plant. By using this NCS, the network-induced delay is analyzed to find the cause of the delay. Furthermore, the fuzzy logic controllers performance is compared with that of conventional proportional-integral-derivative controllers. Based on the experimental results, it is found that the fuzzy logic controller can be a viable choice for an NCS due to its robustness against parameter uncertainty.

   The main concern for this fuzzy approach is with the premise variables for the fuzzy rules. The network-induced delays are normally chosen as the premise variable, and such information has to be available before generating the fuzzy rules. Therefore, only one channel delay, usually sensor-to-controller, can be accommodated in this approach. Different fuzzy rules may also affect the design of the NCS.

NCS is still an open area to which lots of research work is being dedicated. The results presented above just provided some of the techniques that are commonly referred to. It is also worth mentioning that the aforementioned analysis methods are usually specific to the control techniques used and may not be applicable to other control techniques due to different network setups and characteristics (protocols). Furthermore, none of these techniques are perfect because many unrealistic assumptions were used.

1.4 Research Motivation

Key motivations to this book come from several sources. The most general motivation is from the widespread industrial application of NCSs. Even though many efforts have been dedicated to the research of NCSs due to its wide practical use, there are still many open areas to be studied. Furthermore, current techniques for analysis and control of NCSs apply many unrealistic assumptions, which leaves much room for improvements.

A second motivation arises from the nature of the time-delay happens in the communication network, especially in the random access network. We have learnt from the previous section that in most industrial networks, such as CAN, EtherNet™, and Internet, network-induced time-delays are stochastic processes, which are random in nature. This feature ideally corresponds to the Markov process in probability theory. Therefore, the Markov process is believed to be an ideal model of these random time-delays, which provides a more realistic modeling method of the time-delay. Massive research work on the Markov jump linear system that is available in literature also makes this methodology very promising. On the other hand, most research works on time-delay systems by using Lyapunov–Krasovskii technique are with the restrictions on the derivative of the time-delay $\tau(t)$, that is, $\dot{\tau}(t) < 1$. Other assumptions in the research of time-delay systems, such as time-delay is bounded by the maximum sampling interval and sampling intervals are small enough, also introduce conservatism to the results. In this book, we try to relax such restrictions by a novel modeling procedure of NCSs.

Moreover, network-induced time-delays are input delays in the system analysis while so far performance control with disturbance attenuation has not been properly addressed for systems with uncertain time-varying pure input delays. The spirit of the Razumikhin-type method [96] which is an important approach to investigate the delay systems, is based on the use of functions, and it provides methods for determining the stability and solution bounds for retarded functional differential equations. It has been shown recently that the Razumikhin-type method is effective in solving delay problems, especially time-varying delay problems [97]. This leads to our belief that the utilization of the Razumikhin-type method should be effective in analyzing the stability of the disturbance attenuation problem with time-delays.

It should be noted that for systems with time-varying input delays, it is difficult to analyze disturbance attenuation based on the gain characterization, because the state variation depends not only on the current but also the history of exterior disturbance input. We also attempt to solve this problem in this work.

Furthermore, it is also necessary for us to further consider the robust stability against parametric uncertainties in the NCS model. This is because the parametric uncertainties play an important factor responsible for the stability and performance of an NCS. So far, there are very few attempts in the literature. And the existing design methods have not fully addressed the network effects, i.e., network delays and data packet dropout, when analyzing the design problems which makes the results unrealistic. Hence, we need to investigate and find a new technique for designing a robust controller for a class of NCSs in the presence of network effects and parameter uncertainties.

An additional motivation arises from the successful approach of fuzzy logic control (FLC) that overcomes the design problem for nonlinear systems. Due to its complexity natures, controller design for nonlinear systems remains to be an open problem. One of the most common approaches [125] is to linearize the system about an operating point, and uses linear control theory to design a controller. This approach is successful when the operating point of the system is restricted to a certain region. When a wide range operation of the system is required, however, the method may fail. Fuzzy system theory enables us to utilize qualitative, linguistic information about a highly complex nonlinear system to construct a mathematical model for it. Recent studies show that a T-S fuzzy model can be used to approximate global behaviors of a highly complex nonlinear system. Unlike conventional modeling which uses a single model to describe the global behavior of a system, fuzzy modeling is essentially a multi-model approach in which simple sub-models (linear models) are combined to describe the global behavior of the system.

The final and somewhat peripheral motivation is that many control design problems are normally formulated in terms of inequalities rather than simple equalities and a lot of problems in control engineering systems can be formulated as linear matrix inequality (LMI) feasibility problems [98] or BMI problems. Some common convex programming tools, such as ellipsoid methods, interior point methods and methods of alternating convex projections, can be applied to solve the LMIs. However, the interior-point method has been proven to be extremely efficient in solving the LMI with significant computational complexity. The LMI framework provides a tractable method to solve the problems which has either analytical solution or non-analytical solution. Furthermore, a very powerful and efficient toolbox in MATLAB® [99] has been available for solving LMI feasible and optimization problems by interior point methods. Some algorithms have been developed to convert BMI problems, which are non-convex problems that are hard to solve, into iterative LMI problems.

1.5 Contribution of the Book

In this work, our focus is on the control *through* networks, not control *of* networks. Therefore, our design objectives involve the performance and stability of a target NCS rather than of the network. More precisely, it will address the problem of analysis and design of control systems when the communication delays are varying in a random fashion.

Our general contention of this book is that to design an NCS necessitates an integrated approach that combines networking (e.g., network measurement, time-delay, data dropout) with decisions based on sensor data (e.g., sampled-data). Furthermore, it is our conviction that although strategies have to be adapted to the specific application areas, it is possible to develop a general methodology by drawing from the foundation of system theory and of networking.

The focus of this book is to establish novel methodologies for stability analysis, control with disturbance attenuation and fault estimation for a class of linear uncertain NCS with random network-induced communication time-delays. Our methodology is unified in two respects. First, it relies solely on a system-theoretic description of the environment where sensors and actuators are deployed, and thus it is general and encompasses specific applications as particular cases. Second, it strives to optimize directly system-related characteristics, such as stability or performance. These characteristics are influenced by packet losses, delays, and jitter but do not coincide with any one of these network-induced characteristics.

In this work, we first model the network effects, that is, network-induced delay and data packet dropout, both in sensor-to-controller and controller-to-actuator channels, by using two Markov processes. Based on this model, stability analysis, disturbance attenuation and fault estimation problems are considered respectively. It should be noted that for systems with time-varying input delays, it is difficult to analyze disturbance attenuation and fault estimation error based on the gain characterization, because the state variation depends not only on the current but also the history of exterior disturbance input. In this work we introduce a new disturbance attenuation notation for systems with input delays. In the light of such formulation, our object is to design a dynamic delay-dependent state/output feedback controller so that both robust stability and a prescribed disturbance attenuation performance for the closed-loop NCSs are achieved, irrespective of the uncertainties and network-induced effects. A fault estimator for NCSs is also designed to ensure that the fault estimation error is less than some prescribed performance level. In parallel, stability control, disturbance attenuation, and filter estimation problems have also been studied for nonlinear NCSs using T-S fuzzy models.

These results are obtained in the realm of stochastic processes due to the stochastic nature of the time-delays in the network. We therefore adopt Razumikhin-type theorem for stochastic systems and based on a newly adapted Lyapunov–Razumikhin method, the results are given in a form of BMIs. An iterative algorithm is proposed to change this non-convex problem into quasi-convex optimization problems, which can be solved effectively by available mathematical tools.

Finally, to demonstrate the effectiveness and advantages of the proposed design methodologies in this book, some numerical examples are given. The simulation results also show that the proposed design methodologies can achieve the stability requirement or the prescribed performance index.

1.6 Book Outline

The contents of the book are as follows. Chapter 2 describes the model of NCSs to be investigated in this book. The rest of the book is divided into two parts. Part I consists of Chapter 3 to Chapter 7, addressing the controller design and fault estimation problem for linear NCSs. Part II includes Chapter 8 to Chapter 13, dealing with controller design problems for nonlinear NCSs.

Chapter 3 provides a basic idea of the modeling of networked-induced delays using two Markov processes. Based on the Lyapunov–Razumikhin method a novel methodology for designing a mode dependent state feedback controller that stabilizes this class of networked systems is proposed. The existence of such controller is given in terms of the solvability of BMIs, which are solved by a proposed iterative algorithm.

Chapter 4 presents the synthesis design procedure of a robust dynamic output feedback control for linear NCSs. Sufficient conditions to the design of such a controller have been derived. An illustrative example is given along with the theoretical presentation.

In Chapter 5, a generalized disturbance attenuation is introduced. This generalized disturbance attenuation reduces to the standard disturbance attenuation characterized by the L_2 gain when the delay is zero. The resulting delay-dependant controller guarantees both robust stability and a prescribed disturbance attenuation performance for the closed-loop NCSs. Relative results along with demonstrative examples are given to show the effectiveness of the design procedure.

In Chapter 6, a robust filter is designed for a class of linear NCSs to ensure the $\mathscr{L}_2$ gain from an exogenous input to an filter error output is less than or equal to a prescribed value, irrespective of the uncertainties and network-induced effects, i.e., network-induced delays and packet dropouts in the sensor-filter channels. Based on the Lyapunov–Razumikhin method, the existence of a delay-dependent filter is given in terms of the solvability of bilinear matrix inequalities.

In Chapter 7, we apply the same formulation of the generalized disturbance attenuation used in Chapter 5. In light of such formulation, it proposes a robust fault estimator that ensures the fault estimation error is less than prescribed performance level, irrespective of the uncertainties and network-induced effects, i.e., network-induced delays and packet dropouts in communication channels, which are to be modeled by the Markov processes.

Chapter 8 gives preliminary background knowledge of the T-S fuzzy model. In Chapter 9, the design state-feedback controllers for a class of nonlinear uncertain NCSs with sensors and actuators connected to a controller via two communication networks is considered. The nonlinear NCSs are at first represented by a T-S model. A set of linear local controllers for each plant of the T-S model are then designed based on the Lyapunov–Razumikhin method. Sufficient conditions for the existence of a mode-dependent state feedback controller for this class of NCSs are derived.

Chapter 10 deals with the problem of dynamic output feedback controller design for a class of nonlinear uncertain NCSs. The T-S fuzzy model is inherited from the previous chapter. Solutions to the design of such a controller have been derived in terms of BMIs.

Chapter 11 presents procedures for designing a dynamic output feedback controller for a class of nonlinear uncertain NCSs. A numerical example is given along with the theoretical presentation.

Chapter 12 investigates the problem of robust fuzzy filter design for a class of nonlinear NCS, which is a development of the results obtained in Chapter 6. By applying Lyapunov–Razumikhin method, the existence of a delay-dependent filter is given in terms of the solvability of BMIs.

Chapter 13 considers the problem of designing a robust fault estimator for a class of nonlinear uncertain NCSs that ensures the fault estimation error is less than prescribed performance level. A demonstrative example is given to validate results.

Concluding remarks are given and suggestions for future research work are discussed in Chapter 14. Finally, some mathematical background knowledge that is used throughout this research is given in the Appendix.

Chapter 2
Modeling of Networked Control Systems

2.1 Formulation of Networked Control Systems

In this chapter, the modeling procedure of NCSs will be presented. Before proceeding to the modeling procedure, the following assumptions will be used throughout this book:

- The sensor is time-driven: the states of the plant are sampled periodically.
- The controller is event-driven: the control signal is calculated as soon as a new sensor data arrives at the controller.
- The actuator is event-driven: the control signal is applied to the plant as soon as a new controller data arrives at the actuator.

In this book, we will study the NCS of which the generalized plant setup is depicted in Figure 2.1. Plant's measurement signals are denoted as $y(t)$ while input

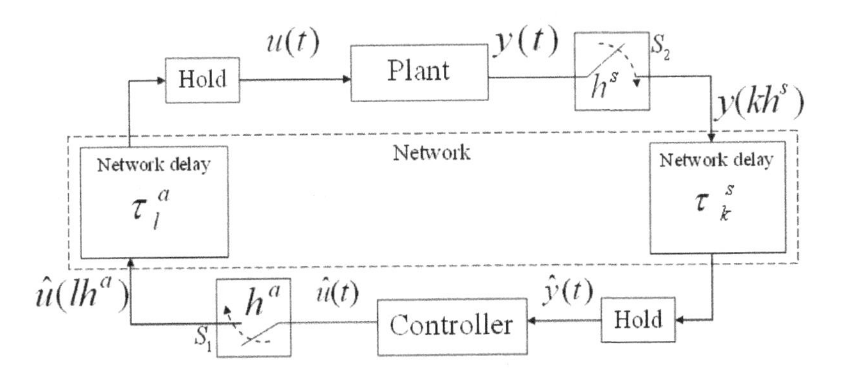

Fig. 2.1 An NCS with random delays

D. Huang and S.K. Nguang: Robust Ctrl. for Uncertain Networked Ctrl. Sys., LNCIS 386, pp. 17–21.
springerlink.com

signals are denoted as $u(t)$. Samplers S_1 and S_2 are time-driven while two ZOHs are event-driven. Furthermore, samplers S_1 and S_2 are not assumed to have the same sampling time. The plant outputs are sampled with a sampling interval h^s and sent through the network at times kh^s, $k \in \mathbb{N}$ while the control signals are sampled with a sampling interval h^a.

As shown in Figure 1.5, when two data messages sent by the sensor side arrive at the controller side during the same sampling time period, only the most recent data packet is used and the previous one is then discarded. This situation is referred to as data packet dropout. Therefore, it is not hard to see that the data packet dropouts can be viewed as prolonged network-induced delays that are at least longer than one sampling period.

In Figure 2.1, it can be noted that the measurement signals $\{y(kh^s),\ k \in \mathbb{N}\}$ are received by the controller side at times $kh^s + \tau_k^s$ where τ_k^s is the delay that the measurement sent at kh^s experiences. When there are data packet dropouts in the communication channel, the signals that the controller receives can be described as follows:

$$\hat{y}(t) = y(kh^s), \forall t \in [kh^s + \tau_k^s, (k+1+n_s)h^s + \tau_{k+n_s+1}^s), \tag{2.1}$$

where $y(kh^s)$ is equal to the last successfully received measurement signal and n_s is the number of consecutive dropouts.

The controller generates control signals using the information of $\hat{y}(t)$. The control signals are then sampled with a sampling interval h^a and sent through the network at times, equal to $\{\hat{u}(lh^a),\ l \in \mathbb{N}\}$. These signals, arrive at the plant side at time $lh^a + \tau_l^a$ where τ_l^a is the delay that the control signal sent at lh^a experiences. Along with network-induced delay, we also include the data packet dropouts into consideration. This leads to the following digital control law:

$$u(t) = \hat{u}(lh^a),\ \forall t \in [lh^a + \tau_l^a, (l+1+n_a)h^a + \tau_{l+n_a+1}^a), \tag{2.2}$$

where n_a is the number of consecutive dropouts.

Now we apply a technique of modeling of continuous-time systems with digital (sampled-data) control in the form of continuous-time systems with delayed control input that was introduced by Mikheev *et al.* [102], Åström and Wittenmark [43], and further developed by Fridman [103, 42]. After minor adaption of the original model, we define:

$$\tau(t) := t - kh^s,\ \forall t \in [kh^s + \tau_k^s, (k+1+n_s)h^s + \tau_{k+n_s+1}^s), \tag{2.3}$$

$$\rho(t) := t - lh^a,\ \forall t \in [lh^a + \tau_l^a, (l+1+n_a)h^a + \tau_{l+n_a+1}^a). \tag{2.4}$$

Based on such a model, (2.1) and (2.2) can be rewritten as:

$$\begin{aligned} \hat{y}(t) &= y(t - \tau(t)), \\ u(t) &= \hat{u}(t - \rho(t)), \end{aligned} \tag{2.5}$$

with

$$\tau(t) \in [\min_k\{\tau_k^s\}, (n^s+1)h^s + \max_k\{\tau_{k+n^s+1}^s\}), \ \forall k \in \mathbb{N}, \tag{2.6}$$

$$\rho(t) \in [\min_l\{\tau_l^a\}, (n^a+1)h^a + \max_l\{\tau_{l+n^a+1}^a\}), \ \forall l \in \mathbb{N}. \tag{2.7}$$

Figure 2.2 shows $\tau(t)$ with respect to time when no data packet dropouts happen, that is, $n_s = 0$, where for all k, $\tau_k^s = \tau^s$ and constant sampling interval h^s with $T = kh^s + \tau^s$. The derivative of $\tau(t)$ is almost always one, except at the sampling times, where $\tau(t)$ drops to τ^s. Hence, the Lyapunov-Krasovskii technique which is with the restrictions on the derivative of the delay $\dot{\tau}(t) < 1$, can not be applied here.

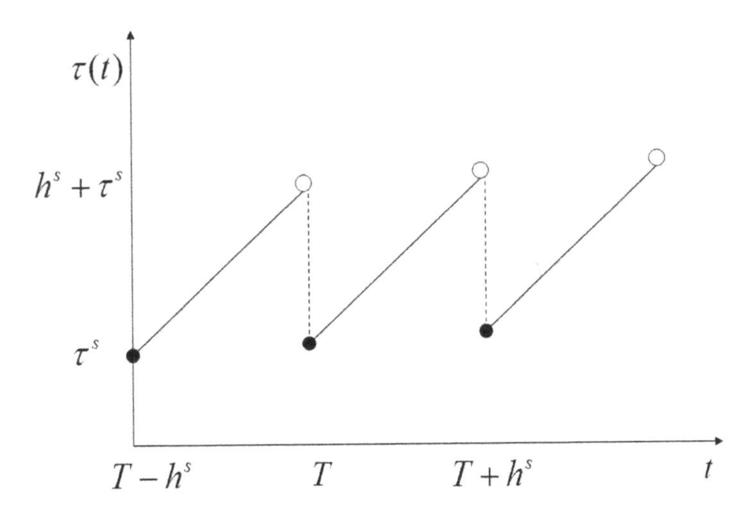

Fig. 2.2 Evolution of $\tau(t)$ with respect to time without packet dropout

Furthermore, if the measurement packet sent at kh^s is drop-out, then $\tau(t)$ increases up to $2h^s + \tau^s$. We can see this scenario from Figure 2.3.

Hence, Figure 2.1 can be reduced to the block diagram depicted in Figure 2.4. This system configuration will be investigated in our book.

2.2 Modeling of Random Network-induced Delays

A stochastic process has the Markov property if the conditional probability distribution of future states of the process, given the present state and all past states, depends only upon the present state and not on any past states, i.e. it is conditionally independent of the past states. In [53], a Markov process is utilized to model these network delays. Modes of the Markov process are defined as different network load conditions. This definition using Markov property is realistic in industry network applications as network traffic and network load conditions are rather of random nature, either in spatial or temporal sense.

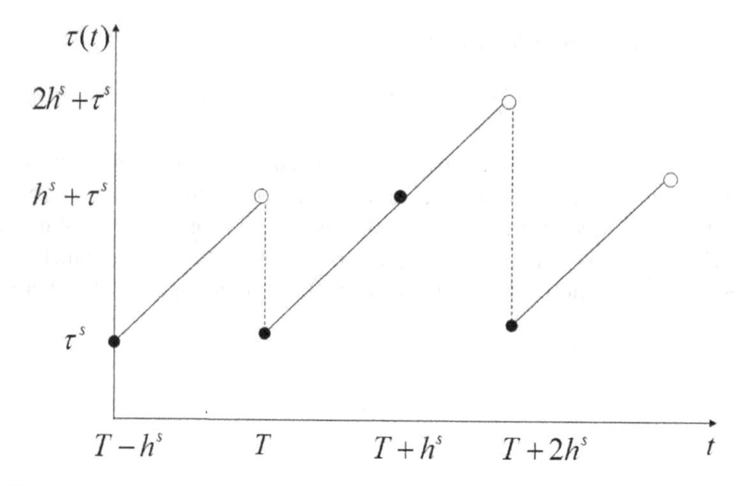

Fig. 2.3 Evolution of $\tau(t)$ with respect to time with packet dropout at kh^s

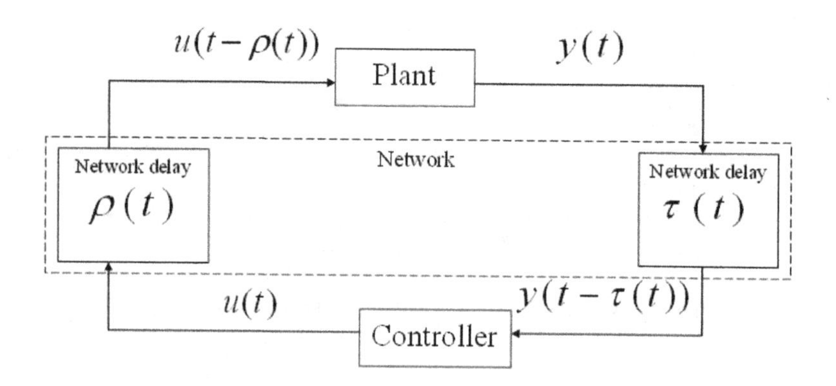

Fig. 2.4 An NCS control system with random delays

In this book, for each mode in the Markov process, a corresponding delay is assumed to be time-varying but upper bounded by a known constant.

Following the same line as [53], we use two Markov processs $\{\eta_1(t)\}$ and $\{\eta_2(t)\}$ to model τ_k^s and τ_l^a, respectively. $\{\eta_1(t)\}$ is a continuous-time discrete-state Markov process taking values in a finite set $\mathscr{S} = \{1, 2, \cdots, s\}$. In some small increment of time from t to $t + \Delta$, the probability that the process makes a transition to some state j, given that it started in some state $i \neq j$ at time t, is given by the following transition probability matrix:

$$Pr\{\eta_1(t+\Delta) = j \mid \eta_1(t) = i\} = \begin{cases} \lambda_{ij}\Delta + o(\Delta), \; i \neq j \\ 1 + \lambda_{ii}\Delta + o(\Delta), \; i = j, \end{cases} \tag{2.8}$$

where $\Delta > 0$, $\lambda_{ij}\Delta < 1$, and $\lim_{\Delta\to 0}\frac{o(\Delta)}{\Delta} = 0$. $o(\Delta)$ represents a quantity that goes to zero faster than Δ as Δ goes to zero. Hence, over a sufficiently small interval of time, the probability of a particular transition is roughly proportional to the duration of that interval. Here $\lambda_{ij} \geq 0$ is the transition rate from mode i to mode j $(i \neq j)$, and $\lambda_{ii} = -\sum_{j=1,j\neq i}^{s}\lambda_{ij}$. $\{\eta_2(t)\}$ takes values in $\mathscr{W} = \{1,2,\cdots,w\}$ with transition probability matrix given by:

$$Pr\{\eta_2(t+\Delta) = l \mid \eta_2(t) = k\} = \begin{cases} \pi_{kl}\Delta + o(\Delta),\ k \neq l \\ 1+\pi_{kk}\Delta + o(\Delta),\ k = l, \end{cases} \tag{2.9}$$

with $\pi_{kl} \geq 0$ and $\pi_{kk} = -\sum_{l=1,l\neq k}^{w}\pi_{kl}$.

In this book, we assume that the mode of the Markov process or state of the network load condition is accessible by the controller and the sensor. The sensor sends the mode of the network load condition and the measurement to the controller. These assumptions are reasonable and they are employed in [53].

Part I
Linear Uncertain Networked Control Systems

Chapter 3
State Feedback Control of Uncertain Networked Control Systems

3.1 Introduction

The Markovian jump linear system (MJLS), which was introduced by Krasovskii and Lidskii [66], is an important class of stochastic dynamic systems that is popular in modeling abrupt changes in the system structure. This is due to the fact that dynamic systems are very often inherently vulnerable to component failures or repairs, sudden environmental disturbances, changing subsystem interconnection and abrupt variations of the operating point of a nonlinear plant etc. This class of systems is normally used to model stochastic processes which change from one mode to another randomly or according to some probabilities. Controllability, stabilizability, observability, and optimal control, as well as some important applications of such systems, can be found in [67, 68, 69, 70, 71, 72, 73, 74] and references therein.

Linear quadratic regulator (LQR) control problems for MJLS in continuous-time domain was addressed in [75, 76]. With the introduction of LMI [98] technology, more results are presented with MJLS and the stability results of MJLS are well established, for instance, in [72, 74, 77]. In addition, $\mathscr{H}_\infty$ control for MJLS were presented in [68, 69, 71, 73]. Guaranteed cost control [78], filtering [79, 71], and fault detection/estimation problems [67, 63] were also among the research areas for MJLS. Some of these results [73, 74] are applied to Markovian jump linear systems with time-delays that are independent of system modes. Markovian jump systems with mode-dependent time-delays have also been studied [77, 78, 71, 72] which provide less conservative results.

According to the characteristics of NCSs, the Markov process is an ideal model of the random time-delays which happen in the random access communication network.

The aim of this chapter is to provide a basic idea of the modeling of NCSs using Markov processes. In this chapter, we consider a class of uncertain NCSs with sensors and actuators connected to a controller via two communication networks in the continuous-time domain. Two Markov processes are used to model the network-induced delays which randomly occurs in both of these two networks. Based on the Lyapunov–Razumikhin method a novel methodology for designing a mode dependent state feedback controller that stabilizes this class of NCSs is proposed. The

D. Huang and S.K. Nguang: Robust Ctrl. for Uncertain Networked Ctrl. Sys., LNCIS 386, pp. 25–35.
springerlink.com

existence of such controller is given in terms of the solvability of BMIs, which are solved by a proposed algorithm in the present work.

3.2 Problem Formulation and Preliminaries

A class of uncertain linear systems under consideration is described by the following equation:

$$\dot{x}(t) = (A + \Delta A)x(t) + (B + \Delta B)u(t), \tag{3.1}$$

where $x(t) \in \mathbb{R}^n$ and $u(t) \in \mathbb{R}^m$ are the plant states and control inputs, respectively. Matrices A and B are known matrices of appropriate dimensions. Matrices ΔA and ΔB characterise the uncertainties in the system and satisfy the following assumption:

Assumption 3.1.

$$[\Delta A \;\; \Delta B] = HF(t)[E_1 \;\; E_2],$$

where H, E_1 and E_2 are known real constant matrices of appropriate dimensions, and $F(t)$ is an unknown matrix function with Lebesgue-measurable elements and satisfies $F(t)^T F(t) \leq I$, in which I is the identity matrix of appropriate dimension.

Following the modeling procedure in Chapter 2, therefore, we have the model of a NCS to be investigated, of which the setup is depicted in Figure 2.4, where $\tau(\eta_1(t),t) \geq 0$ is the sensor-to-controller delay and $\rho(\eta_2(t),t) \geq 0$ is the controller-to-actuator delay. For each mode in the Markov chain, a corresponding delay is assumed to be time-varying but upper bounded by a known constant, that is, $\tau(\eta_1(t),t) \leq \tau^*(i)$ and $\rho(\eta_2(t),t) \leq \rho^*(k)$.

Using the modeling procedure described in Chapter 2, the NCS can be expressed as follows:

$$\text{Plant: } x(t) = (A + \Delta A)x(t) + (B + \Delta B)u(t - \rho(\eta_2(t),t) \tag{3.2}$$

$$\text{Controller: } u(t) = K(\eta_1(t),\eta_2(t))x(t - \tau(\eta_1(t),t)), \tag{3.3}$$

where $K(\eta_1(t),\eta_2(t))$ is the mode-dependent controller gain and yet to be determined.

Substituting (3.3) into (3.2) yields

$$\dot{x}(t) = (A + \Delta A)x(t) + (B + \Delta B)K(\eta_1(t),\eta_2(t))x(t - \tau(\eta_1(t),t) - \rho(\eta_2(t),t)). \tag{3.4}$$

Let $C^{2,1}(\mathbb{R}^n \times \mathscr{S} \times \mathscr{W} \times [-\chi,\infty);\mathbb{R}_+)$ denote the family of all nonnegative functions $V(x(t),\eta_1(t),\eta_2(t),t)$ on $\mathbb{R}^n \times \mathscr{S} \times \mathscr{W} \times [-\chi,\infty)$ which are continuously twice differentiable in x and once differentiable in t. We now cite the the Razumikhin-type theorem established in [105] for the stochastic systems with Markovian jump.

Definition 3.1. Let $\zeta, \alpha_1, \alpha_2$ be all positive numbers and $\delta > 1$. Assume that there exists a function $V \in C^{2,1}(\mathbb{R}^n \times \mathscr{S} \times \mathscr{W} \times [-\chi,\infty);\mathbb{R}_+)$ such that

$$\alpha_1 \|x(t)\|^2 \leq V(x(t),\eta_1(t),\eta_2(t),t) \leq \alpha_2 \|x(t)\|^2,$$

for all $(x(t), \eta_1(t), \eta_2(t), t) \in \mathbb{R}^n \times \mathcal{S} \times \mathcal{W} \times [-\chi, \infty)$, and also for $t \geq 0$,

$$\mathbf{E}\Big[\max_{\eta_1(t)\in\mathcal{S},\eta_2(t)\in\mathcal{W}} \tilde{A}V(x(t),\eta_1(t),\eta_2(t),t)\Big] \leq -\zeta\mathbf{E}\Big[\max_{\eta_1(t)\in\mathcal{S},\eta_2(t)\in\mathcal{W}} V(x(t),\eta_1(t),\eta_2(t),t)\Big], \tag{3.5}$$

provided $x = \{x(\xi) : t - 2\chi \leq \xi \leq t\}$ satisfying:

$$\mathbf{E}\Big[\min_{\eta_1(t)\in\mathcal{S},\eta_2(t)\in\mathcal{W}} V(x(\xi),\eta_1(\xi),\eta_2(\xi),\xi)\Big] < \delta\mathbf{E}\Big[\max_{\eta_1(t)\in\mathcal{S},\eta_2(t)\in\mathcal{W}} V(x(t),\eta_1(t),\eta_2(t),t)\Big], \tag{3.6}$$

for all $t - 2\chi \leq \xi \leq t$. Then the system (3.4) is said to achieve stochastic stability with Markovian jumps.

Here $\tilde{A}V(x(t),\eta_1(t),\eta_2(t),t)$ denotes the weak infinitesimal operator of the random process $V(x(t),\eta_1(t),\eta_2(t),t)$ that is defined as follows:

$$\tilde{A}V(x(t),\eta_1(t),\eta_2(t),t) = \frac{\partial V(\cdot)}{\partial t} + \dot{x}^T(t)\frac{\partial V(\cdot)}{\partial x}\bigg|_{\eta_1=i,\eta_2=k} + \sum_{j=1}^{s}\lambda_{ij}V(x(t),j,k,t) + \sum_{l=1}^{w}\pi_{kl}V(x(t),i,l,t).$$

In this chapter, we also assume $u(t) = 0$ before the first control signal reaches the plant.

For the convenience of notations, $(*)$ is denoted as an ellipsis for terms that are induced by symmetry in the rest of this chapter. $K(\eta_1(t),\eta_2(t))$ is denoted as $K(i,k)$ if $\eta_1(t) = i$ and $\eta_2(t) = k$.

3.3 Main Result

The following theorem provides sufficient conditions for the existence of a mode-dependent state feedback controller for the system (3.4).

Theorem 3.1. *Consider the system (3.4) satisfying Assumption 3.1. If there exist constants $\tau^*(i)$ and $\rho^*(k)$, symmetric matrix $Q(i,k) > 0$, matrix $Y(i,k)$, and positive scalars $\beta_{1_{ik}}$, $\beta_{2_{ik}}$, $\varepsilon_{1_{ik}}$, $\varepsilon_{2_{ik}}$, and ε_{ik} such that the following inequalities hold for all $i \in \mathcal{S}$ and $k \in \mathcal{W}$:*

$$\begin{bmatrix} \begin{pmatrix} Q(i,k)A^T + AQ(i,k) + Y^T(i,k)B^T + BY(i,k) \\ (\tau^*(i) + \rho^*(k))(\beta_{1_{ik}} + 3\beta_{2_{ik}})Q(i,k) \\ +\varepsilon_{ik}HH^T + \lambda_{ii}Q(i,k) + \pi_{kk}Q(i,k) \end{pmatrix} & (*)^T & (*)^T & (*)^T \\ E_1Q(i,k) + E_2Y(i,k) & -\varepsilon_{ik}I & (*)^T & (*)^T \\ S^T(i,k) & 0 & -\mathscr{Q}_1 & (*)^T \\ Z^T(i,k) & 0 & 0 & -\mathscr{Q}_2 \end{bmatrix} < 0 \tag{3.7}$$

$$\begin{bmatrix} -\beta_{1_{ik}}Q(i,k) + \varepsilon_{1_{ik}}HH^T & (*)^T & (*)^T \\ Q(i,k)A^T & -Q(i,k) & (*)^T \\ 0 & E_1Q(i,k) & -\varepsilon_{1_{ik}}I \end{bmatrix} < 0 \tag{3.8}$$

$$\begin{bmatrix} -\beta_{2_{ik}}Q(i,k) + \varepsilon_{2_{ik}}HH^T & (*)^T & (*)^T \\ Y^T(i,k)B^T & -Q(i,k) & (*)^T \\ 0 & E_2Y(i,k) & -\varepsilon_{2_{ik}}I \end{bmatrix} < 0 \tag{3.9}$$

where

$$S(i,k) = [\sqrt{\lambda_{i1}}Q(i,k) \cdots \sqrt{\lambda_{i(i-1)}}Q(i,k)\ \sqrt{\lambda_{i(i+1)}}Q(i,k) \cdots \sqrt{\lambda_{is}}Q(i,k)],$$
$$Z(i,k) = [\sqrt{\pi_{k1}}Q(i,k) \cdots \sqrt{\pi_{k(k-1)}}Q(i,k)\ \sqrt{\pi_{k(k+1)}}Q(i,k) \cdots \sqrt{\pi_{kw}}Q(i,k)],$$

and

$$\mathscr{Q}_1 = diag\{Q(1,k), \cdots, Q(i-1,k), Q(i+1,k), \cdots, Q(s,k)\},$$
$$\mathscr{Q}_2 = diag\{Q(i,1), \cdots, Q(i,k-1), Q(i,k+1), \cdots, Q(i,w)\},$$

with $P(i,k) = Q^{-1}(i,k)$, then system (3.4) is said to achieve stochastic stability via controller (3.3) for all delays $\tau(i,t)$ and $\rho(k,t)$ satisfying

$$0 \le \tau(i,t) + \rho(k,t) \le \tau^*(i) + \rho^*(k),$$

where $K(i,k) = Y(i,k)Q^{-1}(i,k)$.

Proof. Note that for each $\eta_1(t) = i \in \mathscr{S}$ and $\eta_2(t) = k \in \mathscr{W}$,

$$\begin{aligned} & x(t - (\tau(i,t) + \rho(k,t))) \\ &= x(t) - \int_{-(\tau(i,t)+\rho(k,t))}^{0} \dot{x}(t+\theta)d\theta \\ &= x(t) - \int_{-(\tau(i,t)+\rho(k,t))}^{0} [(A + \Delta A)x(t+\theta) \\ &\quad + (B + \Delta B)K(i,k)x(t - (\tau(i,t) + \rho(k,t)) + \theta)]d\theta. \end{aligned} \tag{3.10}$$

Using (3.10), the closed-loop system (3.4) can be rewritten as:

$$\begin{aligned}\dot{x}(t) &= [A+\Delta A+(B+\Delta B)K(i,k)]x(t)\\ &\quad -(B+\Delta B)K(i,k)\int_{-(\tau(i,t)+\rho(k,t))}^{0}[(A+\Delta A)x(t+\theta)\\ &\quad +(B+\Delta B)K(i,k)x(t-(\tau(i,t)+\rho(k,t))+\theta)]d\theta. \end{aligned} \tag{3.11}$$

For the sake of notation simplification, $K(i,k)$ is denoted as K in the rest of this chapter. We also define $\tau(i,t)+\rho(k,t)=\chi(t)$.

Select a stochastic Lyapunov function candidate as

$$V(x(t),\eta_1(t),\eta_2(t),t) = x^T(t)P(\eta_1(t),\eta_2(t))x(t), \tag{3.12}$$

where $P(\eta_1(t),\eta_2(t))$ is the positive symmetric matrix. It follows

$$\alpha_1\|x(t)\|^2 \le V(x(t),\eta_1(t),\eta_2(t),t) \le \alpha_2\|x(t)\|^2, \tag{3.13}$$

where $\alpha_1 = \lambda_{min}(P(\eta_1(t),\eta_2(t)))$ and $\alpha_2 = \lambda_{max}(P(\eta_1(t),\eta_2(t)))$. It is shown that (3.5) is satisfied.

The weak infinitesimal operator $\tilde{A}$ can be considered as the derivative of the function of $V(x(t),\eta_1(t),\eta_2(t),t)$ along the trajectory of the joint Markov process $\{x(t), \eta_1(t), \eta_2(t), t\ge 0\}$ at the point $\{x(t),\eta_1(t)=i,\eta_2(t)=k\}$ at time t; see [110] and [68].

$$\begin{aligned}&\tilde{A}V(x(t),\eta_1(t),\eta_2(t),t)\\ &= \left.\frac{\partial V(\cdot)}{\partial t}+\dot{x}^T(t)\frac{\partial V(\cdot)}{\partial x}\right|_{\eta_1=i,\eta_2=k}+\sum_{j=1}^{s}\lambda_{ij}V(x(t),j,k,t)+\sum_{l=1}^{w}\pi_{kl}V(x(t),i,l,t).\end{aligned} \tag{3.14}$$

It follows from (3.14) that

$$\begin{aligned}&\tilde{A}V(x(t),\eta_1(t),\eta_2(t),t)\\ &= \dot{x}^T(t)P(i,k)x(t)+x^T(t)P(i,k)\dot{x}(t)+\sum_{j=1}^{s}\lambda_{ij}x^T(t)P(j,k)x(t)+\sum_{l=1}^{w}\pi_{kl}x^T(t)P(i,l)x(t)\\ &= x^T(t)[(A+\Delta A)^T P(i,k)+P(i,k)(A+\Delta A)+K^T(B+\Delta B)^T P(i,k)\\ &\quad +P(i,k)(B+\Delta B)K+\sum_{j=1}^{s}\lambda_{ij}P(j,k)+\sum_{l=1}^{w}\pi_{kl}P(i,l)]x(t)\\ &\quad -2\int_{-\chi(t)}^{0}\{x^T(t)P(i,k)(B+\Delta B)K[(A+\Delta A)x(t+\theta)+(B+\Delta B)Kx(t-\chi(t)+\theta)]\}d\theta.\end{aligned}$$

Applying Lemma A.2, we have the following inequalities after some simple algebraic manipulation:

$$
\begin{aligned}
&\tilde{A}V(x(t),\eta_1(t),\eta_2(t),t)\\
&\leq x^T(t)[(A+\Delta A)^T P(i,k)+P(i,k)(A+\Delta A)+K^T(B+\Delta B)^T P(i,k)\\
&\quad +P(i,k)(B+\Delta B)K+\sum_{j=1}^{s}\lambda_{ij}P(j,k)+\sum_{l=1}^{w}\pi_{kl}P(i,l)]x(t)\\
&\quad +\chi(t)[\beta_{1_{ik}}^{-1}x^T(t)P(i,k)(B+\Delta B)K(A+\Delta A)P^{-1}(i,k)\\
&\qquad \times(A+\Delta A)^T K^T(B+\Delta B)^T P(i,k)x(t)\\
&\quad +\beta_{2_{ik}}^{-1}x^T(t)P(i,k)(B+\Delta B)K(B+\Delta B)KP^{-1}(i,k)\\
&\qquad \times K^T(B+\Delta B)^T K^T(B+\Delta B)^T P(i,k)x(t)\\
&\quad +\beta_{1_{ik}}x^T(t+\theta)P(i,k)x(t+\theta)+\beta_{2_{ik}}x^T(t-\chi(t)+\theta)P(i,k)x(t-\chi(t)+\theta)]\\
&=x^T(t)\mathscr{M}_{ik}(\chi(t),\delta)x(t)\\
&\quad +\chi(t)[\beta_{1_{ik}}x^T(t+\theta)P(i,k)x(t+\theta)+\beta_{2_{ik}}x^T(t-\chi(t)+\theta)P(i,k)x(t-\chi(t)+\theta)\\
&\quad -x^T(t)(\beta_{1_{ik}}+\beta_{2_{ik}})\delta P(i,k)x(t)],
\end{aligned}
\tag{3.15}
$$

where $\mathscr{M}_{ik}(\cdot,\cdot)$ is given by:

$$
\begin{aligned}
&\mathscr{M}_{ik}(\chi(t),\delta)\\
&=(A+\Delta A)^T P(i,k)+P(i,k)(A+\Delta A)+K^T(B+\Delta B)^T P(i,k)\\
&\quad +P(i,k)(B+\Delta B)K+\sum_{j=1}^{s}\lambda_{ij}P(j,k)+\sum_{l=1}^{w}\pi_{kl}P(i,l)\\
&\quad +\chi(t)[\beta_{1_{ik}}^{-1}P(i,k)(B+\Delta B)K(A+\Delta A)P^{-1}(i,k)\\
&\qquad \times(A+\Delta A)^T K^T(B+\Delta B)^T P(i,k)\\
&\quad +\beta_{2_{ik}}^{-1}P(i,k)(B+\Delta B)K(B+\Delta B)KP^{-1}(i,k)\\
&\qquad \times K^T(B+\Delta B)^T K^T(B+\Delta B)^T P(i,k)\\
&\quad +(\beta_{1_{ik}}+\beta_{2_{ik}})\delta P(i,k)].
\end{aligned}
\tag{3.16}
$$

Noticing that $\chi(t)=\tau(i,t)+\rho(k,t)$ is upper bounded by $\tau^*(i)+\rho^*(k)$, then

$$\mathscr{M}_{ik}(\chi(t),\delta)\leq\mathscr{M}_{ik}(\tau^*(i)+\rho^*(k),\delta).$$

If (3.8)-(3.9) hold, by applying Lemma A.1 and Schur complement, we get:

$$(A+\Delta A)P^{-1}(i,k)(A+\Delta A)^T<\beta_{1_{ik}}P^{-1}(i,k), \tag{3.17}$$

$$(B+\Delta B)KP^{-1}(i,k)K^T(B+\Delta B)^T<\beta_{2_{ik}}P^{-1}(i,k). \tag{3.18}$$

Using (3.17) and (3.18), $\mathscr{M}_{ik}(\tau^*(i)+\rho^*(k),\delta)$ becomes:

$$
\begin{aligned}
&(A+\Delta A)^T P(i,k)+P(i,k)(A+\Delta A)+K^T(B+\Delta B)^T P(i,k)\\
&+P(i,k)(B+\Delta B)K+\sum_{j=1}^{s}\lambda_{ij}P(j,k)+\sum_{l=1}^{w}\pi_{kl}P(i,l)\\
&+2(\tau^*(i)+\rho^*(k))P(i,k)(B+\Delta B)KP^{-1}(i,k)K^T(B+\Delta B)^T P(i,k)\\
&+(\tau^*(i)+\rho^*(k))(\beta_{1_{ik}}+\beta_{2_{ik}})\delta P(i,k)\\
<\ &(A+\Delta A)^T P(i,k)+P(i,k)(A+\Delta A)+K^T(B+\Delta B)^T P(i,k)\\
&+P(i,k)(B+\Delta B)K+\sum_{j=1}^{s}\lambda_{ij}P(j,k)+\sum_{l=1}^{w}\pi_{kl}P(i,l)\\
&+2(\tau^*(i)+\rho^*(k))\beta_{2_{ik}}P(i,k)+(\tau^*(i)+\rho^*(k))(\beta_{1_{ik}}+\beta_{2_{ik}})\delta P(i,k). \quad (3.19)
\end{aligned}
$$

Hence, if (3.7) holds, it is not hard to see that $\mathscr{M}_{ik}(\tau^*(i)+\rho^*(k),\delta)<0$ for $\delta=1$. Using the continuity property of the eigenvalues of $\mathscr{M}_{ik}(\cdot,\cdot)$ with respect to δ, there exists a $\delta>1$ sufficiently small such that $\mathscr{M}_{ik}(\tau^*(i)+\rho^*(k),1)<0$ still holds.

Thus,

$$
\begin{aligned}
&\tilde{A}V(x(t),\eta_1(t),\eta_2(t),t)\\
&\leq -\alpha x^T(t)x(t)+(\tau^*(i)+\rho^*(k))[\beta_{1_{ik}}x^T(t+\theta)P(i,k)x(t+\theta)\\
&+\beta_{2_{ik}}x^T(t-\chi(t)+\theta)P(i,k)x(t-\chi(t)+\theta)\\
&-(\beta_{1_{ik}}+\beta_{2_{ik}})\delta x^T(t)P(i,k)x(t)], \qquad (3.20)
\end{aligned}
$$

where

$$\alpha=min\{\lambda_{min}(-\mathscr{M}_{ik}(\tau^*(i)+\rho^*(k),1))\}.$$

It is easy to see that $\alpha>0$.

Make use of inequality (3.20) and for any $t\geq 0$ and any $x=\{x(\xi):t-2\chi(t)\leq \xi\leq t\}$ satisfying (3.6), we have

$$\mathbf{E}\Big[\max_{\eta_1(t)\in\mathscr{S},\eta_2(t)\in\mathscr{W}}\tilde{A}V(x(t),\eta_1(t),\eta_2(t),t)\Big]\leq -\alpha\mathbf{E}\Big[\|x(t)\|^2\Big]. \qquad (3.21)$$

Since $\alpha>0$, following (3.13) we can get

$$
\begin{aligned}
&\mathbf{E}\Big[\max_{\eta_1(t)\in\mathscr{S},\eta_2(t)\in\mathscr{W}}\tilde{A}V(x(t),\eta_1(t),\eta_2(t),t)\Big]\\
&\leq -\frac{\alpha}{\alpha_2}\mathbf{E}\Big[\max_{\eta_1(t)\in\mathscr{S},\eta_2(t)\in\mathscr{W}}V(x(t),\eta_1(t),\eta_2(t),t)\Big]. \qquad (3.22)
\end{aligned}
$$

Hence (3.5) is satisfied, which implies that (3.4) is stochastically stable with Markovian jumps.

This completes the proof. □

It should be noted that terms $\beta_{1_{ik}}Q(i,k)$ and $\beta_{2_{ik}}Q(i,k)$ in (3.7)-(3.9) are bilinear, which are difficult to solve. We therefore propose the following iterative algorithm to solve this BMI problem [106].

Algorithm 3.1. *Iterative linear matrix inequality (ILMI) algorithm*

Step 1. Find $Q(i,k)$, $Y(i,k)$ and ε_{ik} such that the following LMIs hold:

$$\begin{bmatrix} \begin{pmatrix} Q(i,k)A^T + AQ(i,k) + Y^T(i,k)B^T + BY(i,k) \\ +\varepsilon_{ik}HH^T + \lambda_{ii}Q(i,k) + \pi_{kk}Q(i,k) \end{pmatrix} & (*)^T & (*)^T & (*)^T \\ E_1 Q(i,k) + E_2 Y(i,k) & -\varepsilon_{ik} I & (*)^T & (*)^T \\ S^T(i,k) & 0 & -\mathscr{Q}_1 & (*)^T \\ Z^T(i,k) & 0 & 0 & -\mathscr{Q}_2 \end{bmatrix} < 0. \tag{3.23}$$

Step 2. For $Q(i,k)$ given in the previous step, find $\beta_{1_{ik}}$, $\beta_{2_{ik}}$, $\varepsilon_{1_{ik}}$, $\varepsilon_{2_{ik}}$, ε_{ik} and $Y(i,k)$ such that the following generalized eigenvalue problem (GEVP) has solutions by replacing $\tau^*(i) + \rho^*(k)$ with τ in (3.7):

$$\max_{\beta_{1_{ik}},\, \beta_{2_{ik}},\, \varepsilon_{1_{ik}},\, \varepsilon_{2_{ik}},\, \varepsilon_{ik},\, Y(i,k)} \tau \text{ s.t. (3.7)-(3.9) hold for } Q(i,k) > 0 \text{ fixed.}$$

Step 3. For $\beta_{1_{ik}}$, $\beta_{2_{ik}}$ and $Y(i,k)$ given in the previous step, find $\varepsilon_{1_{ik}}$, $\varepsilon_{2_{ik}}$, ε_{ik} and $Q(i,k)$ such that the following GEVP has solutions

$$\max_{\varepsilon_{1_{ik}},\, \varepsilon_{2_{ik}},\, \varepsilon_{ik},\, Y(i,k)} \tau \text{ s.t. (3.7)-(3.9) hold for } \beta_{1_{ik}},\ \beta_{2_{ik}} \text{ and } Y(i,k) \text{ fixed.}$$

Step 4. For $Q(i,k)$ and $Y(i,k)$ found in step 2 and step 3, find minimal $\beta^*_{1_{ik}}$ and $\beta^*_{2_{ik}}$ according to the following constraints:

$$\min_{\varepsilon_{1_{ik}}} \beta_{1_{ik}} \text{ s.t. (3.8) holds for } Q(i,k) \text{ fixed,}$$

$$\min_{\varepsilon_{2_{ik}}} \beta_{2_{ik}} \text{ s.t. (3.9) holds for } Q(i,k) \text{ and } Y(i,k) \text{ fixed.}$$

Step 5. $\hbar(i,k)$ is defined as follows:

$$\hbar(i,k) = \frac{\tau \times (\beta_{1_{ik}} + 3\beta_{2_{ik}})}{\beta^*_{1_{ik}} + 3\beta^*_{2_{ik}}}.$$

Step 6. If Return $\hbar(i,k) < \tau^*(i) + \rho^*(k)$, stop. Else, return to step 2.

Remark 3.1. In Step 1, the initial data is obtained by assuming that the system has no time-delay. Note that Steps 2 and 3 are quasi-convex optimization problems [98]. Hence, these two steps guarantee the convergence of τ. In order to reduce the conservatism of imposing a single upper bound τ for all modes, we have Steps 4 and 5. Note that Step 4 is a convex problem.

3.4 Numerical Example

To illustrate the validation of the results obtained previously, we consider the following example taken from [6], where the plant parameters are described as follows:

$$A=\begin{bmatrix}0 & 1\\ 0 & -0.1\end{bmatrix},\ B=\begin{bmatrix}0\\ 0.1\end{bmatrix},\ H=\begin{bmatrix}0.1\\ 0\end{bmatrix},\ E_1=\begin{bmatrix}1 & 0\end{bmatrix},\ E_2=-1.$$

The maximal random time-delays are assumed as follows:

$$\tau^*(1)+\rho^*(1)=0.1,\ \tau^*(1)+\rho^*(2)=0.13,$$
$$\tau^*(2)+\rho^*(1)=0.07,\ \tau^*(2)+\rho^*(2)=0.1.$$

In the following simulation, we assume $F(t) = \sin t$ and it can be seen that $\|F(t)\| \leq 1$. The random time-delays exist in $\mathscr{S} = \{1,2\}$ and $\mathscr{W} = \{1,2\}$, and their transition rate matrices are given by:

$$\Lambda=\begin{bmatrix}-3 & 3\\ 2 & -2\end{bmatrix},\ \Pi=\begin{bmatrix}-1 & 1\\ 2 & -2\end{bmatrix}.$$

Therefore, by applying the algorithm stated in the previous section, we have

$$K(1,1)=\begin{bmatrix}0.2513 & 0.5446\end{bmatrix},\ K(1,2)=\begin{bmatrix}-12.514 & -29.09\end{bmatrix},$$
$$K(2,1)=\begin{bmatrix}-1.4378 & -3.3613\end{bmatrix},\ K(2,2)=\begin{bmatrix}-7.3742 & -17.062\end{bmatrix}.$$

Remark 3.2. In the simulation, we select $F(t) = \sin(10t)$, $\tau(1,t) = \tau(2,t) = 0.025 + 0.025\sin(10t)$, $\rho(1,t) = 0.005+0.005\sin(10t)$ and $\rho(2,t) = 0.025+0.025\sin(10t)$. The state trajectories of the closed-loop system are shown in Figure 3.1 with initial states chosen as $x(0) = x_0 = [1\ 1]^T$. It can be seen that the system is stochastically stable. Figure 3.2 shows the mode transition of the controller during the simulation. The control law $u(t)$ is shown in Figure 3.3. The simulation results demonstrate the validity of the methodology put forward in this chapter. Compared with the results given in [6], our results incorporate the network-induced delays happening in both the sensor-to-controller channel and controller-to-actuator channel. Furthermore, system uncertainties have also been dealt with. The application of Markovian processes also reduces the conservatism.

3.5 Conclusion

In this chapter, a technique of designing a mode-dependent state feedback controller for uncertain NCSs with communication random time-delays has been proposed. The main contribution of this work is that both the sensor-to-controller and controller-to-actuator delays have been taken into account. Two Markov processes have been used to model these two time-delays. The Lyapunov–Razumikhin method has been employed to derive a mode dependent state feedback for this class of systems. Sufficient conditions for the existence of such controller are given in terms of BMIs, which can be solved by the newly proposed algorithm. The effectiveness of this methodology is verified by a numerical example in the last section.

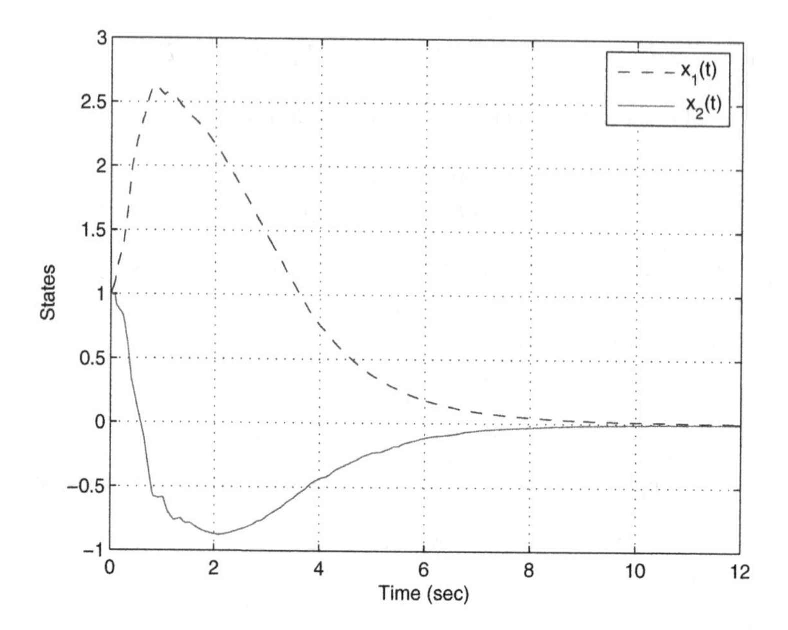

Fig. 3.1 Response of plant states

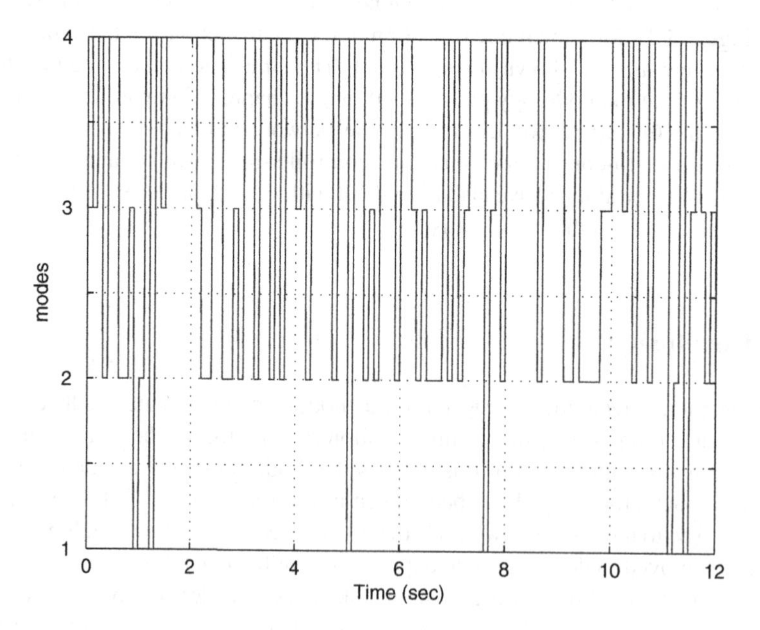

Fig. 3.2 Mode transition

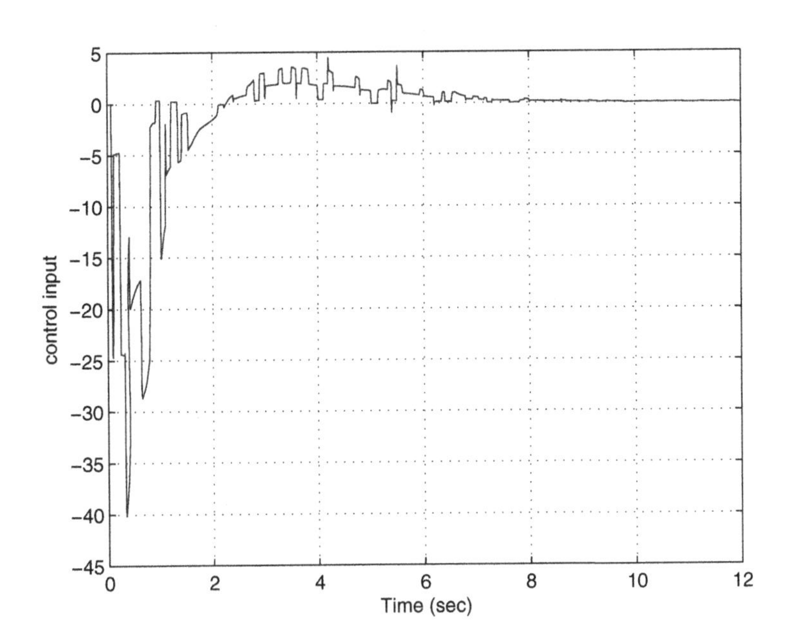

Fig. 3.3 Control input u(t)

Chapter 4
Dynamic Output Feedback Control for Uncertain Networked Control Systems

This chapter investigates the stabilization problem for a class of uncertain NCSs with random communication network-induced delays. The synthesis design procedure of a robust dynamic output feedback controller for linear NCSs is presented in this chapter. The sampling effects and the resulting system delays are incorporated into the design procedure. A system transformation approach is applied to convert the hybrid system which consists of both continuous and discrete signals into a system in the continuous realm. Based on the Lyapunov–Razumikhin method, the existence of such a controller is given in terms of the solvability of BMIs. An iterative algorithm is proposed to change this non-convex problem into quasi-convex optimization problems, which can be solved effectively by available mathematical tools. An illustrative example is given along with the theoretical presentation.

4.1 System Description and Problem Formulation

The uncertain linear system under consideration is assumed to be of the following form:

$$\begin{cases} \dot{x}(t) = (A+\Delta A)x(t) + (B+\Delta B)u(t) \\ y(t) = (C+\Delta C)x(t) \end{cases} \tag{4.1}$$

where $x(t) \in \mathbb{R}^n$, $u(t) \in \mathbb{R}^m$, and $y(t) \in \mathbb{R}^l$ denote plant state, plant output, and control input, respectively. Matrices A , B, and C are of appropriate dimensions. Matrices ΔA, ΔB and ΔC characterize the uncertainties in the system and satisfy the following assumption:

Assumption 4.1.

$$[\Delta A \;\; \Delta B] = H_1 F(t) [E_1 \;\; E_2], \;\; \Delta C = H_2 F(t) E_1,$$

where H_1, H_2, E_1 and E_2 are known real constant matrices of appropriate dimensions, and $F(t)$ is an unknown matrix function with Lebesgue-measurable elements and satisfies $F^T(t)F(t) \leq I$, in which I is the identity matrix of appropriate dimension.

D. Huang and S.K. Nguang: Robust Ctrl. for Uncertain Networked Ctrl. Sys., LNCIS 386, pp. 37–52.
springerlink.com

Following the same modeling procedure as presented in Chapter 2, we can have the overall dynamic output feedback controller and plant models as follows:

$$\text{Controller: } \begin{cases} \dot{\hat{x}}(t) = \hat{A}(\eta_1(t),\eta_2(t))\hat{x}(t) + \hat{B}(\eta_1(t),\eta_2(t))y(t-\tau(\eta_1(t),t)), \\ u(t) = \hat{C}(\eta_1(t),\eta_2(t))\hat{x}(t), \end{cases} \tag{4.2}$$

where $\hat{A}(\eta_1(t),\eta_2(t))$, $\hat{B}(\eta_1(t),\eta_2(t))$, and $\hat{C}(\eta_1(t),\eta_2(t))$ are mode-dependent controller parameters to be designed, and

$$\text{Plant: } \begin{cases} \dot{x}(t) = (A+\Delta A)x(t) + (B+\Delta B)u(t-\rho(\eta_2(t),t)) \\ y(t) = (C+\Delta C)x(t) \end{cases} \tag{4.3}$$

The system setup can be learnt from Figure 2.4.

Therefore, with regard to (4.3) and (4.2), the closed-loop system can be written as:

$$\begin{aligned} \begin{bmatrix} \dot{x}(t) \\ \dot{\hat{x}}(t) \end{bmatrix} &= \begin{bmatrix} A+\Delta A & 0 \\ 0 & \hat{A}(\eta_1(t),\eta_2(t)) \end{bmatrix} \begin{bmatrix} x(t) \\ \hat{x}(t) \end{bmatrix} \\ &+ \begin{bmatrix} 0 & 0 \\ \hat{B}(\eta_1(t),\eta_2(t))(C+\Delta C) & 0 \end{bmatrix} \begin{bmatrix} x(t-\tau(\eta_1(t),t)) \\ \hat{x}(t-\tau(\eta_1(t),t)) \end{bmatrix} \\ &+ \begin{bmatrix} 0 & (B+\Delta B)\hat{C}(\eta_1(t),\eta_2(t)) \\ 0 & 0 \end{bmatrix} \begin{bmatrix} x(t-\rho(\eta_2(t)t)) \\ \hat{x}(t-\rho(\eta_2(t),t)) \end{bmatrix}. \end{aligned} \tag{4.4}$$

Defining $\tilde{x}(t) = [x^T(t) \;\; \hat{x}^T(t)]^T$, (4.4) can be written in the following concise form:

$$\begin{aligned} \dot{\tilde{x}}(t) &= \mathscr{A}(\eta_1(t),\eta_2(t))\tilde{x}(t) + \mathscr{B}(\eta_1(t),\eta_2(t))\tilde{x}(t-\tau(\eta_1(t),t)) \\ &+ \mathscr{C}(\eta_1(t),\eta_2(t))\tilde{x}(t-\rho(\eta_2(t),t)) \end{aligned} \tag{4.5}$$

where

$$\mathscr{A}(\eta_1(t),\eta_2(t)) = \begin{bmatrix} A+\Delta A & 0 \\ 0 & \hat{A}(\eta_1(t),\eta_2(t)) \end{bmatrix},$$

$$\mathscr{B}(\eta_1(t),\eta_2(t)) = \begin{bmatrix} 0 & 0 \\ \hat{B}(\eta_1(t),\eta_2(t))(C+\Delta C) & 0 \end{bmatrix},$$

$$\mathscr{C}(\eta_1(t),\eta_2(t)) = \begin{bmatrix} 0 & (B+\Delta B)\hat{C}(\eta_1(t),\eta_2(t)) \\ 0 & 0 \end{bmatrix}.$$

The aim of this chapter is to design a dynamic output feedback controller of the form (4.2) such that the following inequality holds for $t \geq 0$:

$$\begin{aligned} &\mathbf{E}\Big[\max_{\eta_1(t)\in\mathscr{S},\eta_2(t)\in\mathscr{W}} \tilde{A}V(x(t),\eta_1(t),\eta_2(t),t)\Big] \\ &\leq -\zeta\mathbf{E}\Big[\max_{\eta_1(t)\in\mathscr{S},\eta_2(t)\in\mathscr{W}} V(x(t),\eta_1(t),\eta_2(t),t)\Big], \end{aligned} \tag{4.6}$$

provided $x = \{x(\xi) : t - 2\chi \leq \xi \leq t\}$ satisfying:

$$\mathbf{E}\Big[\min_{\eta_1(t)\in\mathscr{S},\eta_2(t)\in\mathscr{W}} V(x(\xi),\eta_1(\xi),\eta_2(\xi),\xi)\Big] < \delta\mathbf{E}\Big[\max_{\eta_1(t)\in\mathscr{S},\eta_2(t)\in\mathscr{W}} V(x(t),\eta_1(t),\eta_2(t),t)\Big] \quad (4.7)$$

for all $t - 2\chi \leq \xi \leq t$. Then the system (4.5) is said to achieve stochastic stability with Markovian jumps.

In this chapter, we assume $u(t) = 0$ before the first control signal reaches the plant. For notation simplicity, we will denote $\mathscr{A}(\eta_1(t),\eta_2(t)) = \mathscr{A}_{ik}$ when $\eta_1(t) = i \in \mathscr{S}$ and $\eta_2(t) = k \in \mathscr{W}$, and wherever appropriate.

4.2 Main Result

In this chapter, we assume that τ_k^s and τ_l^a are bounded. According to (2.6) and (2.7), there is no loss of generality to assume $\tau(i,t) \leq \tau^*(i)$ and $\rho(k,t) \leq \rho^*(k)$ where $\tau^*(i)$ and $\rho^*(k)$ are known constants. Let us denote the total maximal delay as $\hbar_{ik} = \tau^*(i) + \rho^*(k)$. The following theorem provides sufficient conditions for the existence of a dynamic output feedback controller for the system (4.5).

Theorem 4.1. *Consider the system (4.5) satisfying Assumption 4.1. For given constants $\hbar_{ik}$, ε_1, ε_2, ε_3, ε_4, and ε_5, if there exist symmetric matrices $X(i,k)$, $Y(i,k)$, $R_{1_{ik}}$, $R_{2_{ik}}$, $R_{3_{ik}}$, $R_{4_{ik}}$, and $R_{5_{ik}}$, matrices $F(i,k)$ and $L(i,k)$, and positive scalars β_{1ik}, β_{2ik}, such that the following inequalities hold:*

$$\begin{bmatrix} Y(i,k) & I \\ I & X(i,k) \end{bmatrix} > 0, \quad (4.8)$$

$$\Omega(i,k) < 0, \quad (4.9)$$

$$\Upsilon(i,k) < 0, \quad (4.10)$$

$$\begin{bmatrix} R_{4_{ik}} & (*)^T \\ \Lambda_i^T & \mathscr{Q}_{1_{ik}} \end{bmatrix} > 0, \quad (4.11)$$

$$\begin{bmatrix} R_{5_{ik}} & (*)^T \\ \Pi_k^T & \mathscr{Q}_{2_{ik}} \end{bmatrix} > 0, \quad (4.12)$$

$$\begin{bmatrix} -R_{1_{ik}} & (*)^T & (*)^T & (*)^T \\ 0 & -I & (*)^T & (*)^T \\ 0 & -Y(i,k) & -R_{2_{ik}} & (*)^T \\ 0 & 0 & 0 & -R_{3_{ik}} \end{bmatrix} < 0, \quad (4.13)$$

$$\begin{bmatrix} -\beta_{2_{ik}}Y(i,k) & (*)^T & (*)^T & (*)^T & (*)^T & (*)^T \\ -\beta_{2_{ik}}I & -\beta_{2_{ik}}X(i,k) & (*)^T & (*)^T & (*)^T & (*)^T \\ L^T(i,k)B^T & L^T(i,k)B^T X(i,k) & -Y(i,k) & (*)^T & (*)^T & (*)^T \\ 0 & 0 & -I & -X(i,k) & (*)^T & (*)^T \\ \varepsilon_4 H_1^T & \varepsilon_4 H_1^T X(i,k) & 0 & 0 & -\varepsilon_4 I & (*)^T \\ 0 & 0 & E_2 L(i,k) & 0 & 0 & -\varepsilon_4 I \end{bmatrix} < 0, \quad (4.14)$$

$$\begin{bmatrix} -\beta_{2_{ik}}Y(i,k) & (*)^T & (*)^T & (*)^T & (*)^T & (*)^T \\ -\beta_{2_{ik}}I & -\beta_{2_{ik}}X(i,k) & (*)^T & (*)^T & (*)^T & (*)^T \\ 0 & Y(i,k)C^T F^T(i,k) & -Y(i,k) & (*)^T & (*)^T & (*)^T \\ 0 & 0 & -I & -X(i,k) & (*)^T & (*)^T \\ 0 & \varepsilon_5 H_2^T F^T(i,k) & 0 & 0 & -\varepsilon_5 I & (*)^T \\ 0 & 0 & E_1 Y(i,k) & 0 & 0 & -\varepsilon_5 I \end{bmatrix} < 0, \quad (4.15)$$

where

$$\Omega(i,k) = \left[\begin{matrix} \begin{pmatrix} AY(i,k) + Y(i,k)A^T \\ +BL(i,k) + L^T(i,k)B^T \\ +(\beta_{1_{ik}} + 5\beta_{2_{ik}})\hbar_{ik} Y(i,k) \\ +(\lambda_{ii} + \pi_{kk})Y(i,k) \end{pmatrix} & (*)^T & (*)^T & (*)^T \\ (\beta_{1_{ik}} + 5\beta_{2_{ik}})\hbar_{ik} I & \begin{pmatrix} X(i,k)A + A^T X(i,k) \\ +F(i,k)C + C^T F^T(i,k) \\ +(\beta_{1_{ik}} + 5\beta_{2_{ik}})\hbar_{ik} X(i,k) \\ +\sum_{j=1}^{s} \lambda_{ij} X(j,k) \\ +\sum_{l=1}^{w} \pi_{kl} X(i,l) \end{pmatrix} & (*)^T & (*)^T \\ \varepsilon_1 H_1^T & \varepsilon_1 H_1^T X(i,k) & -\varepsilon_1 I & 0 \\ E_1 Y(i,k) + E_2 L(i,k) & E_1 & 0 & -\varepsilon_1 I \\ 0 & \varepsilon_2 H_2^T F^T(i,k) & 0 & 0 \\ E_1 Y(i,k) & E_1 & 0 & 0 \\ S^T(i,k) & 0 & 0 & 0 \\ Z^T(i,k) & 0 & 0 & 0 \end{matrix} \right.$$

$$\left. \begin{matrix} (*)^T & (*)^T & (*)^T & (*)^T \\ (*)^T & (*)^T & (*)^T & (*)^T \\ (*)^T & (*)^T & (*)^T & (*)^T \\ (*)^T & (*)^T & (*)^T & (*)^T \\ -\varepsilon_2 I & (*)^T & (*)^T & (*)^T \\ 0 & -\varepsilon_2 I & (*)^T & (*)^T \\ 0 & 0 & -\mathscr{Q}_{1_{ik}} & (*)^T \\ 0 & 0 & 0 & -\mathscr{Q}_{2_{ik}} \end{matrix} \right]$$

$$
\Upsilon(i,k) = \begin{bmatrix}
-\beta_{1_{ik}} Y(i,k) + 2R_{1_{ik}} & (*)^T & (*)^T & (*)^T & (*)^T & (*)^T & (*)^T & (*)^T \\
-\beta_{1_{ik}} I & -\beta_{1_{ik}} X(i,k) & (*)^T & (*)^T & (*)^T & (*)^T & (*)^T & (*)^T \\
Y(i,k)A^T & \begin{pmatrix} -A - L^T(i,k)B^T X(i,k) \\ -Y(i,k)C^T F^T(i,k) \\ -(\lambda_{ii} + \pi_{kk})I \end{pmatrix} & -Y(i,k) + 2R_{2_{ik}} & (*)^T & (*)^T & (*)^T & (*)^T & (*)^T \\
A^T & A^T X(i,k) & -I & -X(i,k) + 2R_{3_{ik}} & (*)^T & (*)^T & (*)^T & (*)^T \\
0 & R_{4_{ik}} & 0 & 0 & -I & (*)^T & (*)^T & (*)^T \\
0 & R_{5_{ik}} & 0 & 0 & 0 & -I & (*)^T & (*)^T \\
\varepsilon_3 H_1^T & \varepsilon_3 H_1^T X(i,k) & 0 & 0 & 0 & 0 & -\varepsilon_3 I & (*)^T \\
0 & 0 & E_1 Y(i,k) & E_1 & 0 & 0 & 0 & -\varepsilon_3 I
\end{bmatrix}
$$

and

$$
\begin{aligned}
S(i,k) &= [\sqrt{\lambda_{i1}} Y(i,k) \cdots \sqrt{\lambda_{i(i-1)}} Y(i,k) \ \sqrt{\lambda_{i(i+1)}} Y(i,k) \cdots \sqrt{\lambda_{is}} Y(i,k)], \\
Z(i,k) &= [\sqrt{\pi_{k1}} Y(i,k) \cdots \sqrt{\pi_{k(k-1)}} Y(i,k) \ \sqrt{\pi_{k(k+1)}} Y(i,k) \cdots \sqrt{\pi_{kw}} Y(i,k)], \\
\Lambda_i &= [\sqrt{\lambda_{i1}} I \cdots \sqrt{\lambda_{i(i-1)}} I \ \sqrt{\lambda_{i(i+1)}} I \cdots \sqrt{\lambda_{is}} I], \\
\Pi_k &= [\sqrt{\pi_{k1}} I \cdots \sqrt{\pi_{k(k-1)}} I \ \sqrt{\pi_{k(k+1)}} I \cdots \sqrt{\pi_{kw}} I], \\
\mathcal{Q}_{1_{ik}} &= diag\{Y(1,k), \cdots, Y(i-1,k), Y(i+1,k), \cdots, Y(s,k)\}, \\
\mathcal{Q}_{2_{ik}} &= diag\{Y(i,1), \cdots, Y(i,k-1), Y(i,k+1), \cdots, Y(i,w)\},
\end{aligned}
$$

then the system (4.5) is said to achieve asymptotic stability for all delays $\tau(i,t)$ *and* $\rho(k,t)$ *satisfying* $\tau(i,t) + \rho(k,t) \le \hbar_{ik}$. *Furthermore, the controller is of the form (4.2) with*

$$
\begin{aligned}
\hat{A}_{ik} = {} & (Y^{-1}(i,k) - X(i,k))^{-1}[-A^T - X(i,k)AY(i,k) - F(i,k)CY(i,k) \\
& -X(i,k)BL(i,k) - \sum_{j=1}^{s} \lambda_{ij} Y^{-1}(j,k)Y(i,k) \\
& -\sum_{l=1}^{w} \pi_{kl} Y^{-1}(i,l)Y(i,k)]Y^{-1}(i,k),
\end{aligned} \tag{4.16}
$$

$$
\hat{B}_{ik} = (Y^{-1}(i,k) - X(i,k))^{-1} F(i,k), \tag{4.17}
$$

$$
\hat{C}_{ik} = LY^{-1}(i,k). \tag{4.18}
$$

Proof. Note that for each $\eta_1(t) = i \in \mathscr{S}$ and $\eta_2(t) = k \in \mathscr{W}$ for the system (4.5) at time t, it follows from Leibniz–Newton formula that

$$\begin{aligned}
&\tilde{x}(t-\tau(i,t)) \\
&= \tilde{x}(t) - \int_{-\tau(i,t)}^{0} \dot{\tilde{x}}(t+\theta)d\theta \\
&= \tilde{x}(t) - \int_{-\tau(i,t)}^{0} [\mathscr{A}_{ik}\tilde{x}(t+\theta) + \mathscr{B}_{ik}\tilde{x}(t-\tau(i,t)+\theta) + \mathscr{C}_{ik}\tilde{x}(t-\rho(k,t)+\theta)]d\theta.
\end{aligned}$$

Apply the same transformation to $\tilde{x}(t-\rho(k,t))$, the closed-loop system (4.5) can be rewritten as:

$$\begin{aligned}
&\dot{\tilde{x}}(t) \\
&= \mathscr{D}_{ik}\tilde{x}(t) - \mathscr{C}_{ik}\int_{-\rho(k,t)}^{0} [\mathscr{A}_{ik}\tilde{x}(t+\sigma) + \mathscr{B}_{ik}\tilde{x}(t-\tau(i,t)+\sigma) \\
&\quad + \mathscr{C}_{ik}\tilde{x}(t-\rho(k,t)+\sigma)]d\sigma - \mathscr{B}_{ik}\int_{-\tau(i,t)}^{0} [\mathscr{A}_{ik}\tilde{x}(t+\theta) + \mathscr{B}_{ik}\tilde{x}(t-\tau(i,t)+\theta) \\
&\quad + \mathscr{C}_{ik}\tilde{x}(t-\rho(k,t)+\theta)]d\theta, \qquad (4.19)
\end{aligned}$$

where $\tau(i,t)$ and $\rho(k,t)$ are constant and $\mathscr{D}_{ik} = \mathscr{A}_{ik} + \mathscr{B}_{ik} + \mathscr{C}_{ik}$ for the sake of simplification of notation.

Select a stochastic Lyapunov function candidate as

$$V(\tilde{x}(t), \eta_1(t), \eta_2(t), t) = \tilde{x}^T(t)P(\eta_1(t), \eta_2(t))\tilde{x}(t), \qquad (4.20)$$

where $P(\eta_1(t), \eta_2(t))$ is the positive constant symmetric matrix for each $\eta_1(t) = i \in \mathscr{S}$ and $\eta_2(t) = k \in \mathscr{W}$. It follows

$$\alpha_1\|\tilde{x}(t)\|^2 \le V(\tilde{x}(t), \eta_1(t), \eta_2(t), t) \le \alpha_2\|\tilde{x}(t)\|^2, \qquad (4.21)$$

where $\alpha_1 = \lambda_{min}(P(\eta_1(t), \eta_2(t)))$ and $\alpha_2 = \lambda_{max}(P(\eta_1(t), \eta_2(t)))$.

The weak infinitesimal operator $\tilde{A}$ can be considered as the derivative of the function of $V(x(t), \eta_1(t), \eta_2(t), t)$ along the trajectory of the joint Markov process $\{x(t), \eta_1(t), \eta_2(t), t \ge 0\}$ at the point $\{x(t), \eta_1(t) = i, \eta_2(t) = k\}$ at time t;

$$\begin{aligned}
&\tilde{A}V(x(t), \eta_1(t), \eta_2(t), t) \\
&= \frac{\partial V(\cdot)}{\partial t} + \dot{x}^T(t)\frac{\partial V(\cdot)}{\partial x}\bigg|_{\eta_1=i,\eta_2=k} + \sum_{j=1}^{s}\lambda_{ij}V(x(t), j, k, t) + \sum_{l=1}^{w}\pi_{kl}V(x(t), i, l, t).
\end{aligned} \qquad (4.22)$$

It follows from (4.22) that

$$
\begin{aligned}
&\tilde{A}V(\tilde{x}(t),\eta_1(t),\eta_2(t),t)\\
&= \dot{\tilde{x}}^T(t)P(i,k)\tilde{x}(t)+\tilde{x}^T(t)P(i,k)\dot{\tilde{x}}(t)+\sum_{j=1}^{s}\lambda_{ij}\tilde{x}^T(t)P(j,k)\tilde{x}(t)+\sum_{l=1}^{w}\pi_{kl}\tilde{x}^T(t)P(i,l)\tilde{x}(t)\\
&= \tilde{x}^T(t)[\mathscr{D}_{ik}^T P(i,k)+P(i,k)\mathscr{D}_{ik}]\tilde{x}(t)+\sum_{j=1}^{s}\lambda_{ij}\tilde{x}^T(t)P(j,k)\tilde{x}(t)+\sum_{l=1}^{w}\pi_{kl}\tilde{x}^T(t)P(i,l)\tilde{x}(t)\\
&\quad -2\int_{-\tau(i,t)}^{0}\tilde{x}^T(t)P(i,k)\mathscr{B}_{ik}[\mathscr{A}_{ik}\tilde{x}(t+\theta)+\mathscr{B}_{ik}\tilde{x}(t-\tau(i,t)+\theta)+\mathscr{C}_{ik}\tilde{x}(t-\rho(k,t)+\theta)]d\theta\\
&\quad -2\int_{-\rho(k,t)}^{0}\tilde{x}^T(t)P(i,k)\mathscr{C}_{ik}[\mathscr{A}_{ik}\tilde{x}(t+\sigma)+\mathscr{B}_{ik}\tilde{x}(t-\tau(i,t)+\sigma)+\mathscr{C}_{ik}\tilde{x}(t-\rho(k,t)+\sigma)]d\sigma\\
&\le \tilde{x}^T(t)(\mathscr{D}_{ik}^T P(i,k)+P(i,k)\mathscr{D}_{ik})\tilde{x}(t)+\sum_{j=1}^{s}\lambda_{ij}\tilde{x}^T(t)P(j,k)\tilde{x}(t)+\sum_{l=1}^{w}\pi_{kl}\tilde{x}^T(t)P(i,l)\tilde{x}(t)\\
&\quad +\int_{-\tau(i,t)}^{0}\Big[\frac{1}{\beta_{1_{ik}}}\tilde{x}^T(t)P(i,k)\mathscr{B}_{ik}\mathscr{A}_{ik}P^{-1}(i,k)\mathscr{A}_{ik}^T\mathscr{B}_{ik}^T P(i,k)\tilde{x}(t)+\beta_{1_{ik}}\tilde{x}^T(t+\theta)P(i,k)\tilde{x}(t+\theta)\\
&\quad +\frac{1}{\beta_{2_{ik}}}\tilde{x}^T(t)P(i,k)\mathscr{B}_{ik}\mathscr{B}_{ik}P^{-1}(i,k)\mathscr{B}_{ik}^T P(i,k)\mathscr{B}_{ik}^T\tilde{x}(t)\\
&\quad +\beta_{2_{ik}}\tilde{x}^T(t-\tau(i,t)+\theta)P(i,k)\tilde{x}(t-\tau(i,t)+\theta)\\
&\quad +\frac{1}{\beta_{2_{ik}}}\tilde{x}^T(t)P(i,k)\mathscr{B}_{ik}\mathscr{C}_{ik}P^{-1}(i,k)\mathscr{C}_{ik}^T\mathscr{B}_{ik}^T P(i,k)\tilde{x}(t)\\
&\quad +\beta_{2_{ik}}\tilde{x}^T(t-\rho(k,t)+\theta)P(i,k)\tilde{x}(t-\rho(k,t)+\theta)\Big]\,d\theta\\
&\quad +\int_{-\rho(k,t)}^{0}\Big[\frac{1}{\beta_{1_{ik}}}\tilde{x}^T(t)P(i,k)\mathscr{C}_{ik}\mathscr{A}_{ik}P^{-1}(i,k)\mathscr{A}_{ik}^T\mathscr{C}_{ik}^T P(i,k)\tilde{x}(t)+\beta_{1_{ik}}\tilde{x}^T(t+\sigma)P(i,k)\tilde{x}(t+\sigma)\\
&\quad +\frac{1}{\beta_{2_{ik}}}\tilde{x}^T(t)P(i,k)\mathscr{C}_{ik}\mathscr{B}_{ik}P^{-1}(i,k)\mathscr{B}_{ik}^T\mathscr{C}_{ik}^T P(i,k)\tilde{x}(t)\\
&\quad +\beta_{2_{ik}}\tilde{x}^T(t-\tau(i,t)+\sigma)P(i,k)\tilde{x}(t-\tau(i,t)+\sigma)\\
&\quad +\frac{1}{\beta_{2_{ik}}}\tilde{x}^T(t)P(i,k)\mathscr{C}_{ik}\mathscr{C}_{ik}P^{-1}(i,k)\mathscr{C}_{ik}^T\mathscr{C}_{ik}^T P(i,k)\tilde{x}(t)\\
&\quad +\beta_{2_{ik}}\tilde{x}^T(t-\rho(k,t)+\sigma)P(i,k)\tilde{x}(t-\rho(k,t)+\sigma)\Big]\,d\sigma.
\end{aligned}
$$

Suppose that:

$$\mathscr{A}_{ik}P^{-1}(i,k)\mathscr{A}_{ik}^T < \beta_{1_{ik}}P^{-1}(i,k), \tag{4.23}$$

$$\mathscr{B}_{ik}P^{-1}(i,k)\mathscr{B}_{ik}^T < \beta_{2_{ik}}P^{-1}(i,k), \tag{4.24}$$

$$\mathscr{C}_{ik}P^{-1}(i,k)\mathscr{C}_{ik}^T < \beta_{2_{ik}}P^{-1}(i,k). \tag{4.25}$$

Then we can get:

$$
\begin{aligned}
&\tilde{A}V(\tilde{x}(t),\eta_1(t),\eta_2(t),t)\\
&\leq \tilde{x}^T(t)(\mathscr{D}_{ik}^T P(i,k)+P(i,k)\mathscr{D}_{ik})\tilde{x}(t)+\sum_{j=1}^{s}\lambda_{ij}\tilde{x}^T(t)P(j,k)\tilde{x}(t)+\sum_{l=1}^{w}\pi_{kl}\tilde{x}^T(t)P(i,l)\tilde{x}(t)\\
&\quad +3(\tau(i,t)+\rho(k,t))\beta_{2_{ik}}\tilde{x}^T(t)P(i,k)\tilde{x}(t)+\hbar_{ik}(\beta_{1_{ik}}+2\beta_{2_{ik}})\delta\tilde{x}^T(t)P(i,k)\tilde{x}(t)\\
&\quad +\int_{-\tau(i,t)}^{0}\left[\beta_{1_{ik}}\tilde{x}^T(t+\theta)P(i,k)\tilde{x}(t+\theta)+\beta_{2_{ik}}\tilde{x}^T(t-\tau(i,t)+\theta)P(i,k)\tilde{x}(t-\tau(i,t)+\theta)\right.\\
&\quad \left.+\beta_{2_{ik}}\tilde{x}^T(t-\rho(k,t)+\theta)P(i,k)\tilde{x}(t-\rho(k,t)+\theta)\right]d\theta\\
&\quad +\int_{-\rho(k,t)}^{0}\left[\beta_{1_{ik}}\tilde{x}^T(t+\sigma)P(i,k)\tilde{x}(t+\sigma)+\beta_{2_{ik}}\tilde{x}^T(t-\tau(i,t)+\sigma)P(i,k)\tilde{x}(t-\tau(i,t)+\sigma)\right.\\
&\quad \left.+\beta_{2_{ik}}\tilde{x}^T(t-\rho(k,t)+\sigma)P(i,k)\tilde{x}(t-\rho(k,t)+\sigma)\right]d\sigma\\
&\quad -\hbar_{ik}(\beta_{1_{ik}}+2\beta_{2_{ik}})\delta\tilde{x}^T(t)P(i,k)\tilde{x}(t)\\
&\leq \tilde{x}^T(t)\mathscr{M}_{ik}((\tau(i,t)+\rho(k,t)),\delta)\tilde{x}-\hbar_{ik}(\beta_{1_{ik}}+2\beta_{2_{ik}})\delta\tilde{x}^T(t)P(i,k)\tilde{x}(t)\\
&\quad +\int_{-\tau(i,t)}^{0}\left[\beta_{1_{ik}}\tilde{x}^T(t+\theta)P(i,k)\tilde{x}(t+\theta)+\beta_{2_{ik}}\tilde{x}^T(t-\tau(i,t)+\theta)P(i,k)\tilde{x}(t-\tau(i,t)+\theta)\right.\\
&\quad \left.+\beta_{2_{ik}}\tilde{x}^T(t-\rho(k,t)+\theta)P(i,k)\tilde{x}(t-\rho(k,t)+\theta)\right]d\theta\\
&\quad +\int_{-\rho(k,t)}^{0}\left[\beta_{1_{ik}}\tilde{x}^T(t+\sigma)P(i,k)\tilde{x}(t+\sigma)+\beta_{2_{ik}}\tilde{x}^T(t-\tau(i,t)+\sigma)P(i,k)\tilde{x}(t-\tau(i,t)+\sigma)\right.\\
&\quad \left.+\beta_{2_{ik}}\tilde{x}^T(t-\rho(k,t)+\sigma)P(i,k)\tilde{x}(t-\rho(k,t)+\sigma)\right]d\sigma,
\end{aligned}
$$

where $\mathscr{M}_{ik}(\cdot,\cdot)$ is given by:

$$
\begin{aligned}
&\mathscr{M}_{ik}((\tau(i,t)+\rho(k,t)),\delta)\\
&=\mathscr{D}_{ik}^T P(i,k)+P(i,k)\mathscr{D}_{ik}+3(\tau(i,t)+\rho(k,t))\beta_{2_{ik}}P(i,k)\\
&\quad +\hbar_{ik}(\beta_{1_{ik}}+2\beta_{2_{ik}})\delta P(i,k)+\sum_{j=1}^{s}\lambda_{ij}P(j,k)+\sum_{l=1}^{w}\pi_{kl}P(i,l).
\end{aligned}
$$

In this chapter the time-delays are assumed to be bounded, hence $\tau(i,t)+\rho(k,t)$ can also be assumed to be bounded, that is, $\tau(i,t)+\rho(k,t)\leq\hbar_{ik}$, where $\hbar_{ik}$ is the constant given in the theorem. Using this fact, we learn that

$$\mathscr{M}_{ik}((\tau(i,t)+\rho(k,t)),\delta)\leq\mathscr{M}_{ik}(\hbar_{ik},\delta).$$

Hence, if (4.9) holds, it can be shown later that $\mathscr{M}_{ik}(\hbar_{ik},\delta)<0$ for $\delta=1$. Then we get

$$
\begin{aligned}
&\tilde{A}V(\tilde{x}(t),\eta_1(t),\eta_2(t),t)\\
&<\int_{-\tau(i,t)}^{0}\Big[\beta_{1_{ik}}\tilde{x}^T(t+\theta)P(i,k)\tilde{x}(t+\theta)+\beta_{2_{ik}}\tilde{x}^T(t-\tau(i,t)+\theta)P(i,k)\tilde{x}(t-\tau(i,t)+\theta)\\
&\quad +\beta_{2_{ik}}\tilde{x}^T(t-\rho(k,t)+\theta)P(i,k)\tilde{x}(t-\rho(k,t)+\theta)\Big]d\theta\\
&\quad +\int_{-\rho(k,t)}^{0}\Big[\beta_{1_{ik}}\tilde{x}^T(t+\sigma)P(i,k)\tilde{x}(t+\sigma)+\beta_{2_{ik}}\tilde{x}^T(t-\tau(i,t)+\sigma)P(i,k)\tilde{x}(t-\tau(i,t)+\sigma)\\
&\quad +\beta_{2_{ik}}\tilde{x}^T(t-\rho(k,t)+\sigma)P(i,k)\tilde{x}(t-\rho(k,t)+\sigma)\Big]d\sigma\\
&\quad -\hbar_{ik}(\beta_{1_{ik}}+2\beta_{2_{ik}})\delta\tilde{x}^T(t)P(i,k)\tilde{x}(t)-\alpha\tilde{x}^T(t)\tilde{x}(t),
\end{aligned}
$$

where

$$\alpha = min\{\lambda_{min}(-\mathscr{M}_{ik}(\hbar_{ik},1))\}.$$

It is easy to see that $\alpha > 0$.

Then by Dynkin's formula [70], we have the following result:

$$\begin{aligned}
&\mathbf{E}\{V(x(t),\eta_1(t),\eta_2(t),t)\} - \mathbf{E}\{V(x(0),\eta_1(0),\eta_2(0),0)\} \\
&\leq \int_{-\tau(i,t)}^{0} \Big[\beta_{1_{ik}}\mathbf{E}\{\int_0^{T_f} \tilde{x}^T(t+\theta)P(i,k)\tilde{x}(t+\theta)dt\} + \beta_{2_{ik}}\mathbf{E}\{\int_0^{T_f} \tilde{x}^T(t-\tau(i,t)+\theta)P(i,k) \\
&\times\tilde{x}(t-\tau(i,t)+\theta)dt\} + \beta_{2_{ik}}\mathbf{E}\{\int_0^{T_f} \tilde{x}^T(t-\rho(k,t)+\theta)P(i,k)\tilde{x}(t-\rho(k,t)+\theta)dt\}\Big] d\theta \\
&+ \int_{-\rho(k,t)}^{0} \Big[\beta_{1_{ik}}\mathbf{E}\{\int_0^{T_f} \tilde{x}^T(t+\sigma)P(i,k)\tilde{x}(t+\sigma)dt\} + \beta_{2_{ik}}\mathbf{E}\{\int_0^{T_f} \tilde{x}^T(t-\tau(i,t)+\sigma)P(i,k) \\
&\times\tilde{x}(t-\tau(i,t)+\sigma)dt\} + \beta_{2_{ik}}\mathbf{E}\{\int_0^{T_f} \tilde{x}^T(t-\rho(k,t)+\sigma)P(i,k)\tilde{x}(t-\rho(k,t)+\sigma)dt\}\Big] d\sigma \\
&-\hbar_{ik}(\beta_{1_{ik}} + 2\beta_{2_{ik}})\delta\mathbf{E}\{\int_0^{T_f} \tilde{x}^T(t)P(i,k)\tilde{x}(t)dt\} - \alpha\mathbf{E}\Big[\|x(t)\|^2\Big].
\end{aligned}$$

Now applying Razumikhin-type theorem for stochastic systems with Markovian jumps from [105] (see proof on page 6 in [105]), we know that for $x = \{x(\xi) : t-2\chi \leq \xi \leq t\} \in L^2_{\mathscr{F}_t}([-2\chi,0];\mathbb{R}^n)$ for any $\delta > 1$, the following inequality holds:

$$\begin{aligned}
&\mathbf{E}\Big[\min_{\eta_1(t)\in\mathscr{S},\eta_2(t)\in\mathscr{W}} V(x(\xi),\eta_1(\xi),\eta_2(\xi),\xi)\Big] \\
&< \delta\mathbf{E}\Big[\max_{\eta_1(t)\in\mathscr{S},\eta_2(t)\in\mathscr{W}} V(x(t),\eta_1(t),\eta_2(t),t)\Big].
\end{aligned} \tag{4.26}$$

Since $\alpha > 0$, following (4.21) we can get:

$$\begin{aligned}
&\mathbf{E}\Big[\max_{\eta_1(t)\in\mathscr{S},\eta_2(t)\in\mathscr{W}} \tilde{A}V(x(t),\eta_1(t),\eta_2(t),t)\Big] \\
&\leq -\frac{\alpha}{\alpha_2}\mathbf{E}\Big[\max_{\eta_1(t)\in\mathscr{S},\eta_2(t)\in\mathscr{W}} V(x(t),\eta_1(t),\eta_2(t),t)\Big].
\end{aligned} \tag{4.27}$$

This satisfies (4.6) and we can say the system (4.1) is stochastically stable.

Hereinafter, we will show that (4.9) guarantees $\mathscr{M}_{ik}(\hbar_{ik},1) < 0$.

By using the partition $P(i,k) = \begin{bmatrix} X(i,k) & Y^{-1}(i,k)-X(i,k) \\ Y^{-1}(i,k)-X(i,k) & X(i,k)-Y^{-1}(i,k) \end{bmatrix}$, multiplying each side of $\mathscr{M}_{ik}(\hbar_{ik},1) < 0$ to the left by J_{ik}^T and to the right by J_{ik} where $J_{ik} = \begin{bmatrix} Y(i,k) & I \\ Y(i,k) & 0 \end{bmatrix}$, using Assumption 4.1 and Schur complement, and applying the controllers defined as in (4.16)-(4.18) yields:

$$\begin{bmatrix} \begin{pmatrix} AY(i,k)+Y(i,k)A^T \\ +BL(i,k)+L^T(i,k)B^T \\ +(\beta_{1_{ik}}+5\beta_{2_{ik}})\hbar_{ik}Y(i,k) \\ +(\lambda_{ii}+\pi_{kk})Y(i,k) \end{pmatrix} & (*)^T & (*)^T & (*)^T \\ (\beta_{1_{ik}}+5\beta_{2_{ik}})\hbar_{ik}I & \begin{pmatrix} X(i,k)A+A^TX(i,k) \\ +F(i,k)C+C^TF^T(i,k) \\ +(\beta_{1_{ik}}+5\beta_{2_{ik}})\hbar_{ik}X(i,k) \\ +\sum_{j=1}^{s}\lambda_{ij}X(j,k) \\ +\sum_{l=1}^{w}\pi_{kl}X(i,l) \end{pmatrix} & (*)^T & (*)^T \\ S^T(i,k) & 0 & -\mathscr{Q}_{1_{ik}} & (*)^T \\ Z^T(i,k) & 0 & 0 & -\mathscr{Q}_{2_{ik}} \end{bmatrix}$$
$$+\mathscr{H}_1F(t)\mathscr{E}_1+\mathscr{E}_1^TF^T(t)\mathscr{H}_1^T+\mathscr{H}_2F(t)\mathscr{E}_2+\mathscr{E}_2^TF^T(t)\mathscr{H}_2^T$$
$$<0, \tag{4.28}$$

where

$$\mathscr{H}_1=\begin{bmatrix} H_1 \\ X(i,k)H_1 \\ 0 \\ 0 \end{bmatrix}, \mathscr{H}_2=\begin{bmatrix} 0 \\ F(i,k)H_2 \\ 0 \\ 0 \end{bmatrix},$$
$$\mathscr{E}_1=\begin{bmatrix} E_1Y(i,k)+E_2L(i,k) & E_1 & 0 & 0 \end{bmatrix},$$
$$\mathscr{E}_2=\begin{bmatrix} E_1Y(i,k) & E_1 & 0 & 0 \end{bmatrix}.$$

Using Lemma A.1, it is easy to see that (4.9) guarantees the existence of (4.28), which infers $\mathscr{M}_{ik}(\hbar_{ik},1)<0$. Using the continuity property of the eigenvalues of $\mathscr{M}_{ik}(\cdot,\cdot)$ with respect to δ, there exists a sufficiently small $\varepsilon>0$ such that $\mathscr{M}_{ik}(\hbar_{ik},1+\varepsilon)<0$. Hence, there exists a $\delta>1$ such that $\mathscr{M}_{ik}(\hbar_{ik},\delta)<0$ still holds.

Next, it will be shown that (4.10)-(4.15) are derived from (4.23)-(4.25).

Firstly, the inequality (4.23) can be rewritten as follows by applying Schur complement:

$$\begin{bmatrix} -\beta_{1_{ik}}P^{-1}(i,k) & \mathscr{A}_{ik} \\ \mathscr{A}_{ik}^T & -P(i,k) \end{bmatrix}<0. \tag{4.29}$$

Using Assumption 4.1, multiplying (4.29) to the left by $\begin{bmatrix} J_{ik}^TP(i,k) & 0 \\ 0 & J_{ik}^T \end{bmatrix}$ and to the right by $\begin{bmatrix} P(i,k)J_{ik} & 0 \\ 0 & J_{ik} \end{bmatrix}$, and using the controllers defined as (4.16)-(4.18) yields:

$$
\begin{bmatrix}
-\beta_{1_{ik}} Y(i,k) & (*)^T & (*)^T & (*)^T \\
-\beta_{1_{ik}} I & -\beta_{1_{ik}} X(i,k) & (*)^T & (*)^T \\
Y(i,k)A^T & \begin{pmatrix} -A - L^T(i,k)B^T X(i,k) \\ -Y(i,k)C^T F^T(i,k) \\ -\sum_{j=1}^{s} \lambda_{ij} Y(i,k) Y^{-1}(j,k) \\ -\sum_{l=1}^{w} \pi_{kl} Y(i,k) Y^{-1}(i,l) \end{pmatrix} & -Y(i,k) & (*)^T \\
A^T & A^T X(i,k) & -I & -X(i,k)
\end{bmatrix}
+ \begin{bmatrix} H_1 \\ X(i,k)H_1 \\ 0 \\ 0 \end{bmatrix} F(t) \begin{bmatrix} 0 & 0 & E_1 Y(i,k) & E_1 \end{bmatrix}
+ \begin{bmatrix} 0 & 0 & E_1 Y(i,k) & E_1 \end{bmatrix}^T F^T(t) \begin{bmatrix} H_1 \\ X(i,k)H_1 \\ 0 \\ 0 \end{bmatrix}^T
< 0. \tag{4.30}
$$

To address the term $-\sum_{j=1}^{s} \lambda_{ij} Y(i,k) Y^{-1}(j,k) - \sum_{l=1}^{w} \pi_{kl} Y(i,k) Y^{-1}(i,l)$, we first rewrite (4.30) into the following equivalent form:

$$
\begin{bmatrix}
-\beta_{1_{ik}} Y(i,k) + 2R_{1_{ik}} & (*)^T & (*)^T & (*)^T \\
-\beta_{1_{ik}} I & \begin{pmatrix} -\beta_{1_{ik}} X(i,k) \\ +(\sum_{j=1, j\neq i}^{s} \lambda_{ij} Y^{-1}(j,k))(\sum_{j=1, j\neq i}^{s} \lambda_{ij} Y^{-1}(j,k)) \\ +(\sum_{l=1, l\neq k}^{w} \pi_{kl} Y^{-1}(i,l))(\sum_{l=1, l\neq k}^{w} \pi_{kl} Y^{-1}(i,l)) \end{pmatrix} & (*)^T & (*)^T \\
Y(i,k)A^T & \begin{pmatrix} -A - L^T(i,k)B^T X(i,k) \\ -Y(i,k)C^T F^T(i,k) - (\lambda_{ii} + \pi_{kk}) I \end{pmatrix} & -Y(i,k) + 2R_{2_{ik}} & (*)^T \\
A^T & A^T X(i,k) & -I & -X(i,k) + 2R_{3_{ik}}
\end{bmatrix}
$$

$$
+ \begin{bmatrix}
-R_{1_{ik}} & (*)^T & (*)^T & (*)^T \\
0 & -(\sum_{j=1, j\neq i}^{s} \lambda_{ij} Y^{-1}(j,k))(\sum_{j=1, j\neq i}^{s} \lambda_{ij} Y^{-1}(j,k)) & (*)^T & (*)^T \\
0 & -\sum_{j=1, j\neq i}^{s} \lambda_{ij} Y(i,k) Y^{-1}(j,k) & -R_{2_{ik}} & (*)^T \\
0 & 0 & 0 & -R_{3_{ik}}
\end{bmatrix}
$$

$$
+ \begin{bmatrix}
-R_{1_{ik}} & (*)^T & (*)^T & (*)^T \\
0 & -(\sum_{l=1, l\neq k}^{w} \pi_{kl} Y^{-1}(i,l))(\sum_{l=1, l\neq k}^{w} \pi_{kl} Y^{-1}(i,l)) & (*)^T & (*)^T \\
0 & -\sum_{l=1, l\neq k}^{w} \pi_{kl} Y(i,k) Y^{-1}(i,l) & -R_{2_{ik}} & (*)^T \\
0 & 0 & 0 & -R_{3_{ik}}
\end{bmatrix}
$$

$$
+\begin{bmatrix} H_1 \\ X(i,k)H_1 \\ 0 \\ 0 \end{bmatrix} F(t)\begin{bmatrix} 0 & 0 & E_1Y(i,k) & E_1 \end{bmatrix}
$$

$$
+\begin{bmatrix} 0 & 0 & E_1Y(i,k) & E_1 \end{bmatrix}^T F^T(t)\begin{bmatrix} H_1 \\ X(i,k)H_1 \\ 0 \\ 0 \end{bmatrix}^T
$$

$$
< 0. \tag{4.31}
$$

On the left hand side of (4.31), if the second and third term are less than zero, we get:

$$
\begin{bmatrix}
-\beta_{1_{ik}}Y(i,k)+2R_{1_{ik}} & (*)^T & (*)^T & (*)^T \\
-\beta_{1_{ik}}I & \begin{pmatrix} -\beta_{1_{ik}}X(i,k) \\ +(\Sigma_{j=1,j\neq i}^{s}\lambda_{ij}Y^{-1}(j,k))(\Sigma_{j=1,j\neq i}^{s}\lambda_{ij}Y^{-1}(j,k)) \\ +(\Sigma_{l=1,l\neq k}^{w}\pi_{kl}Y^{-1}(i,l))(\Sigma_{l=1,l\neq k}^{w}\pi_{kl}Y^{-1}(i,l)) \end{pmatrix} & (*)^T & (*)^T \\
Y(i,k)A^T & \begin{pmatrix} -A-L^T(i,k)B^TX(i,k) \\ -Y(i,k)C^TF^T(i,k)-(\lambda_{ii}+\pi_{kk})I \end{pmatrix} & -Y(i,k)+2R_{2_{ik}} & (*)^T \\
A^T & A^TX(i,k) & -I & -X(i,k)+2R_{3_{ik}}
\end{bmatrix}
$$

$$
+\begin{bmatrix} H_1 \\ X(i,k)H_1 \\ 0 \\ 0 \end{bmatrix} F(t)\begin{bmatrix} 0 & 0 & E_1Y(i,k) & E_1 \end{bmatrix}
$$

$$
+\begin{bmatrix} 0 & 0 & E_1Y(i,k) & E_1 \end{bmatrix}^T F^T(t)\begin{bmatrix} H_1 \\ X(i,k)H_1 \\ 0 \\ 0 \end{bmatrix}^T
$$

$$
< 0. \tag{4.32}
$$

By defining new variables $R_{4_{ik}}$ and $R_{5_{ik}}$ and using (4.11) and (4.12), we get $R_{4_{ik}} > \Sigma_{j=1,j\neq i}^{s}\lambda_{ij}Y^{-1}(j,k)$ and $R_{5_{ik}} > \Sigma_{l=1,l\neq k}^{w}\pi_{kl}Y^{-1}(i,l)$, which also implies that

$$
R_{4_{ik}}R_{4_{ik}} > \left[\sum_{j=1,j\neq i}^{s}\lambda_{ij}Y^{-1}(j,k)\right]\left[\sum_{j=1,j\neq i}^{s}\lambda_{ij}Y^{-1}(j,k)\right],
$$

$$
R_{5_{ik}}R_{5_{ik}} > \left[\sum_{l=1,l\neq k}^{w}\pi_{kl}Y^{-1}(i,l)\right]\left[\sum_{l=1,l\neq k}^{w}\pi_{kl}Y^{-1}(i,l)\right].
$$

Therefore, by applying Lemma A.1 and Schur complement, it is not hard to see that if (4.10) holds, (4.32) is guaranteed and (4.23) is thereby satisfied.

Furthermore, we address the negativeness of the second and third term on the left hand side of (4.31). Firstly, we want the second term less than zero, that is:

$$\begin{bmatrix} -R_{1_{ik}} & (*)^T & (*)^T & (*)^T \\ 0 & -(\Sigma_{j=1,j\neq i}^{s}\lambda_{ij}Y^{-1}(j,k))(\Sigma_{j=1,j\neq i}^{s}\lambda_{ij}Y^{-1}(j,k)) & (*)^T & (*)^T \\ 0 & -\Sigma_{j=1,j\neq i}^{s}\lambda_{ij}Y(i,k)Y^{-1}(j,k) & -R_{2_{ik}} & (*)^T \\ 0 & 0 & 0 & -R_{3_{ik}} \end{bmatrix} < 0. \tag{4.33}$$

By multiplying (4.33) both sides by $\begin{bmatrix} I & 0 & 0 & 0 \\ 0 & (\Sigma_{j=1,j\neq i}^{s}\lambda_{ij}Y^{-1}(j,k))^{-1} & 0 & 0 \\ 0 & 0 & I & 0 \\ 0 & 0 & 0 & I \end{bmatrix}$, we can see that if (4.13) exists, (4.33) holds.

It is straightforward that if (4.13) holds, the third term is negative as well.

(4.14)-(4.15) can be derived from (4.24)-(4.25) using the same procedure.

Besides, $P > 0$ is equivalent to

$$J_{ik}^T P(i,k) J_{ik} = \begin{bmatrix} Y(i,k) & I \\ I & X(i,k) \end{bmatrix} > 0. \tag{4.34}$$

We therefore have the inequality condition (4.8).

We now have the completion of the proof. □

It should be noted that the terms $\beta_{1ik}X(i,k)$ and $\beta_{1ik}Y(i,k)$ in (4.9)-(4.14) are non-convex constraints, which are difficult to solve. The iterative algorithm proposed in the previous chapter is therefore used to change this non-convex problem into quasi-convex optimization problems, which can be solved effectively by available mathematical tools.

4.3 Numerical Example

Consider the following example taken from [26], where the plant parameters are described:

$$A = \begin{bmatrix} -1.7 & 3.8 \\ -1 & 1.8 \end{bmatrix}, B = \begin{bmatrix} 5 \\ 2.01 \end{bmatrix}, C = \begin{bmatrix} 10.1 & 4.5 \end{bmatrix},$$
$$H = \begin{bmatrix} 0.01 \\ 0 \end{bmatrix}, E_1 = \begin{bmatrix} 1 & 0 \end{bmatrix}, E_2 = -1.$$

In our simulation, we assume that the sampling period is 0.05 for both sensor and actuation channels, that is, $h^a = h^s = 0.05$ and $n^s = n^a = 0$, which means no data packet dropout happens in the communication channel. From this, it is not hard to see that the longer the sampling period is or the more data packets are lost, the smaller time-delay the communication channel can tolerate. ε_1, ε_2, ε_3, ε_4, and ε_5 are set to be equivalent to 1.

Furthermore, we assume that the sensor-to-controller communication delay is $|\tau_k^s| < 0.06$, while the controller-to-actuator delay is $|\tau_l^a| < 0.08$. Therefore, by (2.6) and (2.7) we can have $\hbar = 0.24$.

By applying Theorem 4.1 and the algorithm in the previous section, we get the following controller gains after the calculation of (4.16)-(4.18):

$$\hat{A}_{11} = \begin{bmatrix} -2.2664 & 1.7952 \\ -1.5970 & 0.8094 \end{bmatrix}, \hat{B}_{11} = \begin{bmatrix} 0.0587 \\ 0.0614 \end{bmatrix}, \hat{C}_{11} = \begin{bmatrix} 0.0043 & -0.3470 \end{bmatrix},$$
$$\hat{A}_{12} = \begin{bmatrix} -2.7655 & 2.2231 \\ -1.8741 & 0.4015 \end{bmatrix}, \hat{B}_{12} = \begin{bmatrix} 0.0369 \\ 0.0874 \end{bmatrix}, \hat{C}_{12} = \begin{bmatrix} 0.0021 & -0.2541 \end{bmatrix},$$
$$\hat{A}_{21} = \begin{bmatrix} -3.0154 & 1.0121 \\ -0.9847 & 0.7845 \end{bmatrix}, \hat{B}_{21} = \begin{bmatrix} 0.0525 \\ 0.043 \end{bmatrix}, \hat{C}_{21} = \begin{bmatrix} 0.0004 & -0.4313 \end{bmatrix},$$
$$\hat{A}_{22} = \begin{bmatrix} -4.871 & 2.5847 \\ -2.9847 & 1.2584 \end{bmatrix}, \hat{B}_{22} = \begin{bmatrix} 0.0245 \\ 0.0789 \end{bmatrix}, \hat{C}_{22} = \begin{bmatrix} 0.0102 & -0.4432 \end{bmatrix},$$

Figure 4.1 shows the simulation result of state response with $x_0 = [3.5 \ -2.1]$, while the digital control input is plotted in Figure 4.2.

Furthermore, under the same assumptions on the sampling period h^a and h^s, we choose $n^s = 3$ and $n^a = 1$ to model the data dropouts in the communication channel. In this case, sensor-to-controller delays and controller-to-actuator delays are under the same bound as in the previous case, and hence we can have $\hbar = 0.44$. The relative simulation results are shown in Figure 4.3 and Figure 4.4.

These results demonstrate the validity of the methodology put forward in this chapter.

4.4 Conclusion

In this chapter, a technique of designing a dynamic output feedback controller for an uncertain NCS with random communication network-induced delays and data packet dropouts has been proposed. The main contribution of this work is that both the sensor-to-controller and controller-to-actuator delays/dropouts have been taken into account. The Lyapunov–Razumikhin method has been employed to derive such a controller for this class of systems. Sufficient conditions for the existence of such a controller for this class of NCSs are derived. We finally use a numerical example to demonstrate the effectiveness of this methodology in the last section.

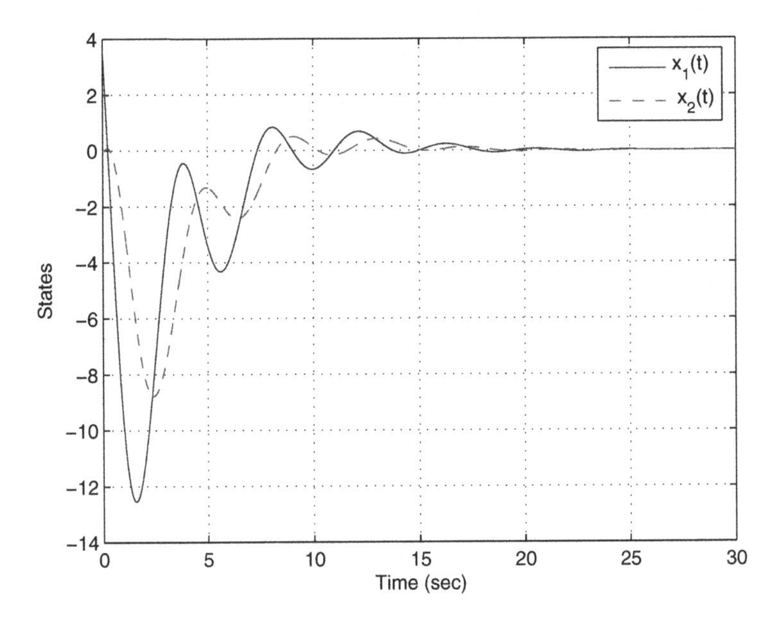

Fig. 4.1 Response of plant states without data packet dropout

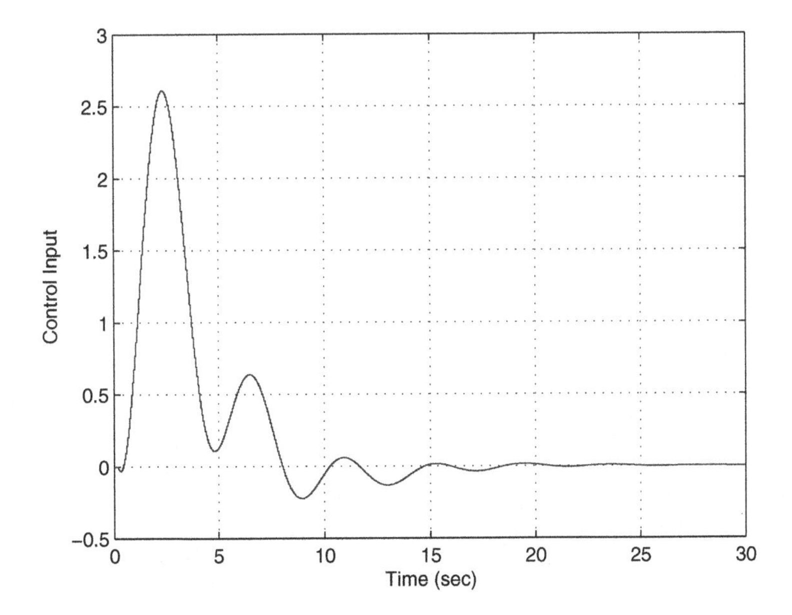

Fig. 4.2 Control input without data packet dropout

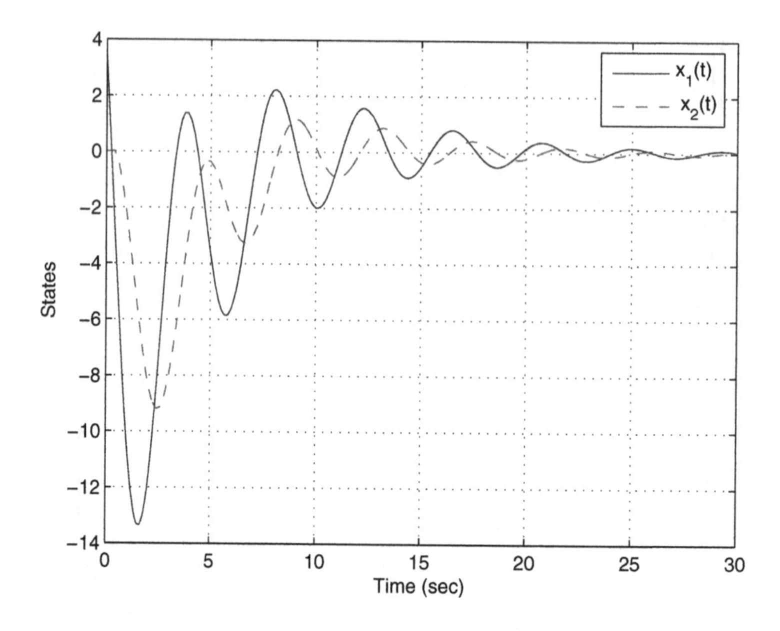

Fig. 4.3 Response of plant states with data packet dropout

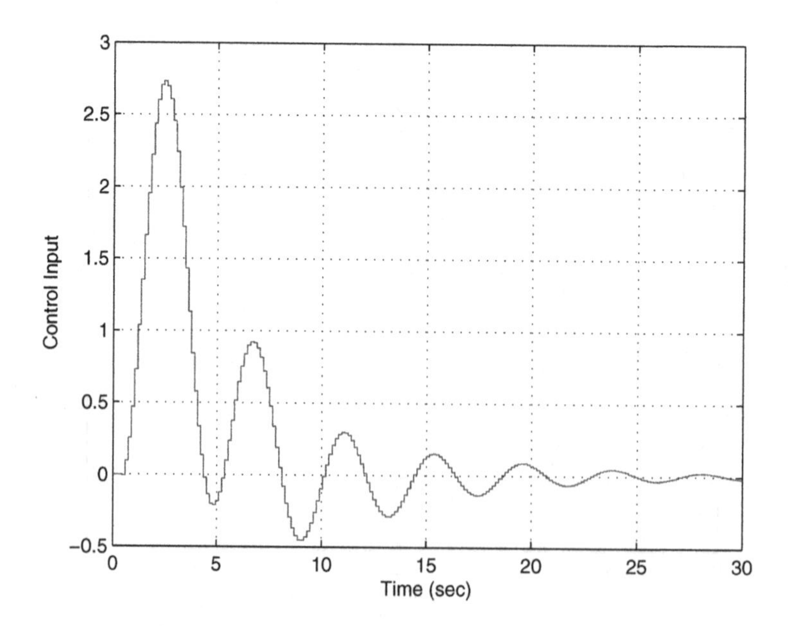

Fig. 4.4 Control input with data packet dropout

Chapter 5
Robust Disturbance Attenuation for Uncertain Networked Control Systems

5.1 Introduction

For time-delay problems encountered in engineering, two approaches are employed. One is to obtain information on the time-delay and subsequently to use this information to solve the problem. The other is to attenuate the effects caused by delay disturbances when the delays cannot be effectively used or obtained. Furthermore, the problem of performance control with disturbance attenuation for time-delay systems has gathered much attention in recent years [73, 80, 81].

The $\mathscr{H}_\infty$ control problem is able to address the issue of system parameter uncertainty, and also be applied to the typical problem of disturbance input control. It was initially formulated [82] in the early 1980s [83] where the $\mathscr{H}_\infty$ norm plays an important role and resulted from the requirement of disturbance attenuation characterized by the $\mathscr{L}_2$ gain. The effectiveness of the controller in attenuating according to the $\mathscr{H}_\infty$ norm has been widely reported and intensively studied for systems without input delays.

Recently, there have been interesting studies investigating the design of the $\mathscr{H}_\infty$ controller to guarantee not only asymptotic stability but also the $\mathscr{H}_\infty$ norm bound of a closed-loop system with time-delays. Based on the solution of a Riccati-like equation, a method to obtain the gain matrix of the $\mathscr{H}_\infty$ controller of linear systems with constant delays was proposed in [84]. In [85], the authors consider the $\mathscr{H}_\infty$ controller design problem for linear systems with time-varying delays in states. In [87] the robust $\mathscr{H}_\infty$ performance for linear delay differential systems is studied with an uncertain constant time-delay and time-varying norm-bounded parameter uncertainties.

The aforementioned results are mostly obtained for systems with state delays. Due to the characteristics of communication networks, network-induced time-delays are input delays. So far, performance control with disturbance attenuation has not been addressed for systems with uncertain time-varying pure input delays.

We attempt to solve this problem in this chapter. It should be noted that for systems with time-varying input delays, it is difficult to analyze disturbance attenuation based on the gain characterization, because the state variation depends not only on

D. Huang and S.K. Nguang: Robust Ctrl. for Uncertain Networked Ctrl. Sys., LNCIS 386, pp. 53–63.
springerlink.com

the current but also the history of exterior disturbance input. In this chapter, a generalized disturbance attenuation will be introduced. This generalized disturbance attenuation reduces to the standard disturbance attenuation characterized by the L_2 gain when the delay is zero. In the light of such formulation, our object is to design a dynamic output feedback controller such that both robust stability and a prescribed disturbance attenuation performance for the closed-loop NCS are achieved, irrespective of the uncertainties and network-induced effects, i.e., network-induced delays and packet dropouts in both the sensor-controller and controller-actuator channels. Based on the Lyapunov–Razumikhin method, the existence of a delay-dependent controller is given in terms of the solvability of BMIs.

5.2 System Description and Problem Formulation

Assume that the uncertain linear continuous state-space model of the plant dynamics is described by the following equations:

$$\begin{cases} \dot{x}(t) = (A+\Delta A)x(t) + (B_1+\Delta B_1)w(t) + (B_2+\Delta B_2)u(t) \\ z(t) = (C_1+\Delta C_1)x(t) + (D_1+\Delta D_1)u(t) \\ y(t) = (C_2+\Delta C_2)x(t) + (D_2+\Delta D_2)w(t) \end{cases} \tag{5.1}$$

where $x(t) \in \mathbb{R}^n$ is the state vector, $u(t) \in \mathbb{R}^m$ is the control input, $w(t) \in \mathbb{R}^p$ is the exogenous disturbance input and/or measurement noise, $y(t) \in \mathbb{R}^l$ and $z(t) \in \mathbb{R}^s$ denote the measurement and regulated output respectively. Matrices A, B_1, B_2, C_1, C_2, D_1, and D_2 are of appropriate dimensions.

Matrices ΔA, ΔB_1, ΔB_2, ΔC_1, ΔC_2, ΔD_1, and ΔD_2 characterize the uncertainties in the system and satisfy the following assumption:

Assumption 5.1.

$$\begin{aligned} \begin{bmatrix} \Delta A & \Delta B_1 & \Delta B_2 \end{bmatrix} &= H_1 F(t) \begin{bmatrix} E_1 & E_2 & E_3 \end{bmatrix}, \\ \begin{bmatrix} \Delta C_1 & \Delta D_1 \end{bmatrix} &= H_2 F(t) \begin{bmatrix} E_1 & E_3 \end{bmatrix}, \\ \begin{bmatrix} \Delta C_2 & \Delta D_2 \end{bmatrix} &= H_3 F(t) \begin{bmatrix} E_1 & E_2 \end{bmatrix}, \end{aligned}$$

where H_1, H_2, H_3, E_1, E_2, and E_3 are known real constant matrices of appropriate dimensions, and $F(t)$ is an unknown matrix function with Lebesgue-measurable elements and satisfies $F(t)^T F(t) \leq I$, in which I is the identity matrix of appropriate dimension.

The overall system setup to be investigated is depicted in Figure 5.1. Following the same lines in Chapter 2 with regard to the modeling of NCSs, a dynamic output feedback controller is constructed at time t as follows:

$$\text{Controller } \mathscr{G}: \begin{cases} \dot{\hat{x}}(t) = \hat{A}(\eta_1(t),\eta_2(t))\hat{x}(t) + \hat{B}(\eta_1(t),\eta_2(t))y(t-\tau(\eta_1(t),t)), \\ u(t) = \hat{C}(\eta_1(t),\eta_2(t))\hat{x}(t), \end{cases} \tag{5.2}$$

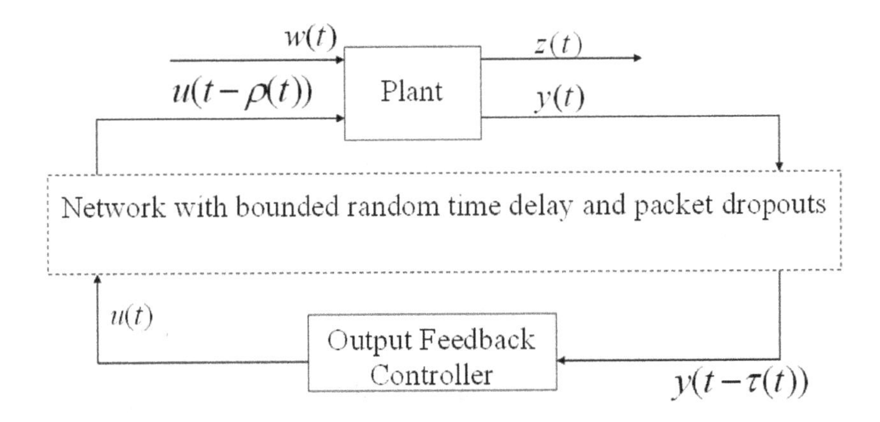

Fig. 5.1 An NCS with random delays, packet dropouts, and disturbance input

where $\hat{A}(\eta_1(t),\eta_2(t))$, $\hat{B}(\eta_1(t),\eta_2(t))$, and $\hat{C}(\eta_1(t),\eta_2(t))$ are mode-dependent controller parameters to be designed. The plant model can be rewritten as:

$$\text{Plant:}\begin{cases} \dot{x}(t) = (A+\Delta A)x(t) + (B_1+\Delta B_1)w(t) + (B_2+\Delta B_2)u(t-\rho(\eta_2(t),t)) \\ z(t) = (C_1+\Delta C_1)x(t) + (D_1+\Delta D_1)u(t-\rho(\eta_2(t),t)) \\ y(t) = (C_2+\Delta C_2)x(t) + (D_2+\Delta D_2)w(t) \end{cases} \tag{5.3}$$

Therefore, with regard to (5.3) and (5.2), the closed-loop system can be written as:

$$\begin{aligned}\begin{bmatrix}\dot{x}(t)\\ \dot{\hat{x}}(t)\end{bmatrix} &= \begin{bmatrix}A+\Delta A & 0\\ 0 & \hat{A}(\eta_1(t),\eta_2(t))\end{bmatrix}\begin{bmatrix}x(t)\\ \hat{x}(t)\end{bmatrix} + \begin{bmatrix}B_1+\Delta B_1\\ 0\end{bmatrix}w(t)\\ &+\begin{bmatrix}0\\ \hat{B}(\eta_1(t),\eta_2(t))(D_2+\Delta D_2)\end{bmatrix}w(t-\tau(\eta_1(t),t))\\ &+\begin{bmatrix}0 & (B_2+\Delta B_2)\hat{C}(\eta_1(t),\eta_2(t))\\ 0 & 0\end{bmatrix}\begin{bmatrix}x(t-\rho(\eta_2(t),t))\\ \hat{x}(t-\rho(\eta_2(t),t))\end{bmatrix}\\ &+\begin{bmatrix}0 & 0\\ \hat{B}(\eta_1(t),\eta_2(t))(C_2+\Delta C_2) & 0\end{bmatrix}\begin{bmatrix}x(t-\tau(\eta_1(t),t))\\ \hat{x}(t-\tau(\eta_1(t),t))\end{bmatrix}. \end{aligned} \tag{5.4}$$

Define $\tilde{x}(t) = [x^T(t)\ \ \hat{x}^T(t)]^T$, $w(t-\tau(\eta_1(t),t)) = v(t)$, and $\omega(t) = [w^T(t)\ \ v^T(t)]^T$. Hence (5.4) can be written in the following concise form :

$$\begin{aligned}\dot{\tilde{x}}(t) &= \mathscr{A}(\eta_1(t),\eta_2(t))\tilde{x}(t) + \mathscr{B}(\eta_1(t),\eta_2(t))\omega(t)\\ &\quad +\mathscr{C}_1(\eta_1(t),\eta_2(t))\tilde{x}(t-\rho(\eta_2(t),t)) + \mathscr{C}_2(\eta_1(t),\eta_2(t))\tilde{x}(t-\tau(\eta_1(t),t)),\end{aligned} \tag{5.5}$$

where

$$\mathscr{A}(\eta_1(t),\eta_2(t)) = \begin{bmatrix} A+\Delta A & 0 \\ 0 & \hat{A}(\eta_1(t),\eta_2(t)) \end{bmatrix},$$

$$\mathscr{B}(\eta_1(t),\eta_2(t)) = \begin{bmatrix} B_1+\Delta B_1 & 0 \\ 0 & \hat{B}(\eta_1(t),\eta_2(t))(D_2+\Delta D_2) \end{bmatrix},$$

$$\mathscr{C}_1(\eta_1(t),\eta_2(t)) = \begin{bmatrix} 0 & (B_2+\Delta B_2)\hat{C}(\eta_1(t),\eta_2(t)) \\ 0 & 0 \end{bmatrix},$$

$$\mathscr{C}_2(\eta_1(t),\eta_2(t)) = \begin{bmatrix} 0 & 0 \\ \hat{B}(\eta_1(t),\eta_2(t))(C_2+\Delta C_2) & 0 \end{bmatrix}.$$

Let $C^{2,1}(\mathbb{R}^n \times \mathscr{S} \times \mathscr{W} \times [-\tau,\infty);\mathbb{R}_+)$ denote the family of all nonnegative functions $V(x(t),\eta_1(t),\eta_2(t),t)$ on $\mathbb{R}^n \times \mathscr{S} \times \mathscr{W} \times [-\tau,\infty)$ which are continuously twice differentiable in x and once differentiable in t. Also, let $\tau > 0$ and $C([-\tau,0];\mathbb{R}^n)$ denote the family of continuous function φ from $[-\tau,0]$ to $\mathbb{R}^n$ with the norm $\| \varphi \| = \sup_{-\tau\leq\theta\leq 0} | \varphi(\theta) |$. Let $(\Omega,\mathscr{F},\{\mathscr{F}_t\}_{t\geq 0},P)$ be a complete probability space with a filtration $\{\mathscr{F}_t\}_{t\geq 0}$ satisfying the usual conditions. Denote by $L^2_{\mathscr{F}_t}([-\tau,0];\mathbb{R}^n)$ the family of all $\{\mathscr{F}_t\}$-measurable $C([-\tau,0];\mathbb{R}^n)$-valued random variables $\phi = \{\phi(\theta) : -\tau \leq \theta \leq 0\}$ such that $\sup_{-\tau\leq\theta\leq 0} \mathbf{E} | \phi(\theta) |^2 < \infty$.

We now cite the the Razumikhin-type theorem established in [105] for the stochastic systems with Markovian jump.

Definition 5.1. Let ζ,α_1,α_2 be all positive numbers and $\delta > 1$. Assume that there exists a function $V \in C^{2,1}(\mathbb{R}^n \times \mathscr{S} \times \mathscr{W} \times [-\chi,\infty);\mathbb{R}_+)$ such that

$$\alpha_1\|x(t)\|^2 \leq V(x(t),\eta_1(t),\eta_2(t),t) \leq \alpha_2\|x(t)\|^2, \tag{5.6}$$

for all $(x(t),\eta_1(t),\eta_2(t),t) \in \mathbb{R}^n \times \mathscr{S} \times \mathscr{W} \times [-\chi,\infty)$, and also for system (5.1), if its zero state response ($x(\phi) = 0$, $\omega(\phi) = 0$, $-\chi \leq \phi \leq 0$) satisfies,

$$\mathbf{E}\left[\int_0^{T_f} z^T(t)z(t)dt\right] \leq \gamma^2\mathbf{E}\left[\int_0^{T_f} \sup_{-\chi\leq\phi\leq 0} \omega^T(t+\phi)\omega(t+\phi)dt\right], \tag{5.7}$$

for any nonzero $\omega(t) \in \mathscr{L}_2[0,\ T_f]$ and $T_f \geq 0$, provided $x = \{x(\xi) : t-2\chi \leq \xi \leq t\} \in L^2_{\mathscr{F}_t}([-2\chi,0];\mathbb{R}^n)$ satisfying:

$$\begin{aligned} &\mathbf{E}\left[\min_{\eta_1(t)\in\mathscr{S},\eta_2(t)\in\mathscr{W}} V(x(\xi),\eta_1(\xi),\eta_2(\xi),\xi)\right] \\ &< \delta\mathbf{E}\left[\max_{\eta_1(t)\in\mathscr{S},\eta_2(t)\in\mathscr{W}} V(x(t),\eta_1(t),\eta_2(t),t)\right], \end{aligned} \tag{5.8}$$

for all $t-2\chi \leq \xi \leq t$. Then the system (5.1) is said to be stochastically stabilizable with a disturbance attenuation level γ.

Remark 5.1. From Definition 5.1, it is easy to find that once there is no time-delay in the system, i.e., $\phi = 0$, (5.7) reduces to $\mathbf{E}\left[\int_0^{T_f} z^T(t)z(t)dt\right] \leq \gamma^2\mathbf{E}\left[\int_0^{T_f} \omega^T(t)\omega(t)dt\right]$, which is $\mathscr{H}_\infty$ control problem.

In this chapter, we assume $u(t) = 0$ before the first control signal reaches the plant. For notation simplicity, we will denote $\mathscr{A}(\eta_1(t), \eta_2(t)) = \mathscr{A}_{ik}$ when $\eta_1(t) = i \in \mathscr{S}$ and $\eta_2(t) = k \in \mathscr{W}$, and wherever appropriate.

5.3 Main Result

In this chapter, we assume that τ_k^s and τ_l^a are bounded. For each $\eta_1(t) = i \in \mathscr{S}$ and $\eta_2(t) = k \in \mathscr{W}$, there is no loss of generality to assume $\tau(i,t) \leq \tau^*(i)$ and $\rho(k,t) \leq \rho^*(k)$. Let us denote the total maximal delay $\hbar_{ik} = \tau^*(i) + \rho^*(k)$. The following theorem provides sufficient conditions for the existence of a dynamic output feedback controller for the system (5.5) that satisfies requirements for robust attenuation with stability.

Theorem 5.1. *Consider the system (5.5) satisfying Assumption 5.1. For the given positive delay-free attenuation constant γ_{d_f}, positive constants $\hbar_{ik}$, ε_1, ε_2, ε_3, ε_4, ε_5, ε_6,, ε_7, ε_8, and ε_9, if there exist symmetric positive matrices $X(i,k)$, $Y(i,k)$, $R_{1_{ik}}$, $R_{2_{ik}}$, $R_{3_{ik}}$, $R_{4_{ik}}$, and $R_{5_{ik}}$, and matrices $F(i,k)$ and $L(i,k)$, and positive scalars $\beta_{1_{ik}}$, $\beta_{2_{ik}}$, such that the following inequalities hold where $i \in \mathscr{S}$ and $k \in \mathscr{W}$:*

$$\begin{bmatrix} Y(i,k) & I \\ I & X(i,k) \end{bmatrix} > 0, \tag{5.9}$$

$$\Upsilon(i,k) < 0, \tag{5.10}$$

$$\Phi(i,k) < 0, \tag{5.11}$$

$$\begin{bmatrix} R_{4_{ik}} & (*)^T \\ \Lambda_i^T & \mathscr{Q}_{1_{ik}} \end{bmatrix} > 0, \tag{5.12}$$

$$\begin{bmatrix} R_{5_{ik}} & (*)^T \\ \Pi_k^T & \mathscr{Q}_{2_{ik}} \end{bmatrix} > 0, \tag{5.13}$$

$$\begin{bmatrix} -R_{1_{ik}} & (*)^T & (*)^T & (*)^T \\ 0 & -I & (*)^T & (*)^T \\ 0 & -Y(i,k) & -R_{2_{ik}} & (*)^T \\ 0 & 0 & 0 & -R_{3_{ik}} \end{bmatrix} < 0, \tag{5.14}$$

$$\begin{bmatrix} -\beta_{2_{ik}} Y(i,k) & (*)^T & (*)^T & (*)^T & (*)^T & (*)^T \\ -\beta_{2_{ik}} I & -\beta_{2_{ik}} X(i,k) & (*)^T & (*)^T & (*)^T & (*)^T \\ L^T(i,k)B_2^T & L^T(i,k)B_2^T X(i,k) & -Y(i,k) & (*)^T & (*)^T & (*)^T \\ 0 & 0 & -I & -X(i,k) & (*)^T & (*)^T \\ \varepsilon_6 H_1^T & \varepsilon_6 H_1^T X(i,k) & 0 & 0 & -\varepsilon_6 I & (*)^T \\ 0 & 0 & E_3 L(i,k) & 0 & 0 & -\varepsilon_6 I \end{bmatrix} < 0, \tag{5.15}$$

$$\begin{bmatrix} -\beta_{2_{ik}}Y(i,k) & (*)^T & (*)^T & (*)^T & (*)^T & (*)^T \\ -\beta_{2_{ik}}I & -\beta_{2_{ik}}X(i,k) & (*)^T & (*)^T & (*)^T & (*)^T \\ 0 & Y(i,k)C_2^T F^T(i,k) & -Y(i,k) & (*)^T & (*)^T & (*)^T \\ 0 & 0 & -I & -X(i,k) & (*)^T & (*)^T \\ 0 & \varepsilon_7 H_3^T F^T(i,k) & 0 & 0 & -\varepsilon_7 I & (*)^T \\ 0 & 0 & E_1 Y(i,k) & 0 & 0 & -\varepsilon_7 I \end{bmatrix} < 0, \quad (5.16)$$

$$\begin{bmatrix} -Y(i,k) & (*)^T & (*)^T & (*)^T & (*)^T & (*)^T & (*)^T & (*)^T \\ -I & -X(i,k) & (*)^T & (*)^T & (*)^T & (*)^T & (*)^T & (*)^T \\ B_1^T & B_1^T X(i,k) & -I & (*)^T & (*)^T & (*)^T & (*)^T & (*)^T \\ 0 & D_2^T F^T(i,k) & 0 & -I & (*)^T & (*)^T & (*)^T & (*)^T \\ \varepsilon_8 H_1^T & \varepsilon_8 H_1^T X(i,k) & 0 & 0 & -\varepsilon_8 I & (*)^T & (*)^T & (*)^T \\ 0 & 0 & E_2 & 0 & 0 & -\varepsilon_8 I & (*)^T & (*)^T \\ 0 & \varepsilon_9 H_3^T F^T(i,k) & 0 & 0 & 0 & 0 & -\varepsilon_9 I & (*)^T \\ 0 & 0 & 0 & E_2 & 0 & 0 & 0 & -\varepsilon_9 I \end{bmatrix} < 0, \quad (5.17)$$

where

$$\Upsilon(i,k) = \left[\begin{array}{cccccccccccccccc} \Xi_1(i,k) & (*)^T & (*)^T & (*)^T & (*)^T & (*)^T & (*)^T & (*)^T & (*)^T & (*)^T & (*)^T & (*)^T & (*)^T & (*)^T & (*)^T \\ (\beta_{1_{ik}} + 6\beta_{2_{ik}})\hbar_{ik} I & \Xi_2(i,k) & (*)^T & (*)^T & (*)^T & (*)^T & (*)^T & (*)^T & (*)^T & (*)^T & (*)^T & (*)^T & (*)^T & (*)^T & (*)^T \\ B_1^T & B_1^T X(i,k) & -\gamma_{d_f} I & (*)^T & (*)^T & (*)^T & (*)^T & (*)^T & (*)^T & (*)^T & (*)^T & (*)^T & (*)^T & (*)^T & (*)^T \\ 0 & D_2^T F^T(i,k) & 0 & -\gamma_{d_f} I & (*)^T & (*)^T & (*)^T & (*)^T & (*)^T & (*)^T & (*)^T & (*)^T & (*)^T & (*)^T & (*)^T \\ C_1 Y(i,k) + D_1 L(i,k) & C_1 & 0 & 0 & -I & (*)^T & (*)^T & (*)^T & (*)^T & (*)^T & (*)^T & (*)^T & (*)^T & (*)^T & (*)^T \\ E_1 Y(i,k) + E_3 L(i,k) & 0 & E_2 & 0 & 0 & -\varepsilon_1 I & (*)^T & (*)^T & (*)^T & (*)^T & (*)^T & (*)^T & (*)^T & (*)^T & (*)^T \\ \varepsilon_1 H_1^T & 0 & 0 & 0 & 0 & 0 & -\varepsilon_1 I & (*)^T & (*)^T & (*)^T & (*)^T & (*)^T & (*)^T & (*)^T & (*)^T \\ E_1 Y(i,k) & E_1 & E_2 & 0 & 0 & 0 & 0 & -\varepsilon_2 I & (*)^T & (*)^T & (*)^T & (*)^T & (*)^T & (*)^T & (*)^T \\ 0 & \varepsilon_2 H_1^T X(i,k) & 0 & 0 & 0 & 0 & 0 & 0 & -\varepsilon_2 I & (*)^T & (*)^T & (*)^T & (*)^T & (*)^T & (*)^T \\ 0 & E_1 & 0 & E_2 & 0 & 0 & 0 & 0 & 0 & -\varepsilon_3 I & (*)^T & (*)^T & (*)^T & (*)^T & (*)^T \\ 0 & \varepsilon_3 H_3^T F^T(i,k) & 0 & 0 & 0 & 0 & 0 & 0 & 0 & 0 & -\varepsilon_3 I & (*)^T & (*)^T & (*)^T & (*)^T \\ E_1 Y(i,k) + E_3 L(i,k) & E_1 & 0 & 0 & 0 & 0 & 0 & 0 & 0 & 0 & 0 & -\varepsilon_4 I & (*)^T & (*)^T & (*)^T \\ 0 & 0 & 0 & 0 & \varepsilon_4 H_2^T & 0 & 0 & 0 & 0 & 0 & 0 & 0 & -\varepsilon_4 I & (*)^T & (*)^T \\ S^T(i,k) & 0 & 0 & 0 & 0 & 0 & 0 & 0 & 0 & 0 & 0 & 0 & 0 & -\mathscr{Q}_{1_{ik}} & (*)^T \\ Z^T(i,k) & 0 & 0 & 0 & 0 & 0 & 0 & 0 & 0 & 0 & 0 & 0 & 0 & 0 & -\mathscr{Q}_{2_{ik}} \end{array}\right]$$

$$\Phi(i,k) = \begin{bmatrix} -\beta_{1_{ik}}Y(i,k)+2R_{1_{ik}} & (*)^T & (*)^T \\ -\beta_{1_{ik}}I & -\beta_{1_{ik}}X(i,k) & (*)^T \\ Y(i,k)A^T & \begin{pmatrix} -A-L^T(i,k)B_2^TX(i,k) \\ -Y(i,k)C_2^TF^T(i,k)-(\lambda_{ii}+\pi_{kk})I \end{pmatrix} & -Y(i,k)+2R_{2_{ik}} \\ A^T & A^TX(i,k) & -I \\ 0 & R_{4_{ik}} & 0 \\ 0 & R_{5_{ik}} & 0 \\ \varepsilon_5 H_1^T & \varepsilon_5 H_1^T X(i,k) & 0 \\ 0 & 0 & E_1 Y(i,k) \end{bmatrix.$$

$$\left.\begin{matrix} (*)^T & (*)^T & (*)^T & (*)^T & (*)^T \\ (*)^T & (*)^T & (*)^T & (*)^T & (*)^T \\ (*)^T & (*)^T & (*)^T & (*)^T & (*)^T \\ -X(i,k)+2R_{3_{ik}} & (*)^T & (*)^T & (*)^T & (*)^T \\ 0 & -I & (*)^T & (*)^T & (*)^T \\ 0 & 0 & -I & (*)^T & (*)^T \\ 0 & 0 & 0 & -\varepsilon_5 I & (*)^T \\ E_1 & 0 & 0 & 0 & -\varepsilon_5 I \end{matrix}\right],$$

and

$$\begin{aligned} \Xi_1(i,k) &= AY(i,k)+Y(i,k)A^T+B_2L(i,k)+L^T(i,k)B_2^T \\ &\quad +(\beta_{1_{ik}}+6\beta_{2_{ik}})\hbar_{ik}Y(i,k)+(\lambda_{ii}+\pi_{kk})Y(i,k), \\ \Xi_2(i,k) &= X(i,k)A+A^TX(i,k)+F(i,k)C_2+C_2^TF^T(i,k) \\ &\quad +(\beta_{1_{ik}}+6\beta_{2_{ik}})\hbar_{ik}X(i,k)+\sum_{j=1}^{s}\lambda_{ij}X(j,k)+\sum_{l=1}^{w}\pi_{kl}X(i,l), \end{aligned}$$

with

$$\begin{aligned} S(i,k) &= [\sqrt{\lambda_{i1}}Y(i,k)\cdots\sqrt{\lambda_{i(i-1)}}Y(i,k)\ \sqrt{\lambda_{i(i+1)}}Y(i,k)\cdots\sqrt{\lambda_{is}}Y(i,k)], \\ Z(i,k) &= [\sqrt{\pi_{k1}}Y(i,k)\cdots\sqrt{\pi_{k(k-1)}}Y(i,k)\ \sqrt{\pi_{k(k+1)}}Y(i,k)\cdots\sqrt{\pi_{kw}}Y(i,k)], \\ \Lambda_i &= [\sqrt{\lambda_{i1}}I\cdots\sqrt{\lambda_{i(i-1)}}I\ \sqrt{\lambda_{i(i+1)}}I\cdots\sqrt{\lambda_{is}}I], \\ \Pi_k &= [\sqrt{\pi_{k1}}I\cdots\sqrt{\pi_{k(k-1)}}I\ \sqrt{\pi_{k(k+1)}}I\cdots\sqrt{\pi_{kw}}I], \\ \mathcal{Q}_{1_{ik}} &= diag\{Y(1,k),\cdots,Y(i-1,k),Y(i+1,k),\cdots,Y(s,k)\}, \\ \mathcal{Q}_{2_{ik}} &= diag\{Y(i,1),\cdots,Y(i,k-1),Y(i,k+1),\cdots,Y(i,w)\}, \end{aligned}$$

then (5.7) holds for all delays $\tau(i,t)$ and $\rho(k,t)$ satisfying $\tau(i,t)+\rho(k,t) \le \hbar_{ik}$ with $\gamma^2 = \gamma_{d_f} + \max(\hbar_{ik})$. Furthermore, the mode dependant controller $\mathcal{G}_{ik}$ is of the form (5.2) with

$$\hat{A}_{ik} = [Y^{-1}(i,k) - X(i,k)]^{-1}[-A^T - X(i,k)AY(i,k) - F(i,k)C_2Y(i,k)$$
$$-X(i,k)B_2L(i,k) - \sum_{j=1}^{s} \lambda_{ij}Y^{-1}(j,k)Y(i,k)$$
$$-\sum_{l=1}^{w} \pi_{kl}Y^{-1}(i,l)Y(i,k)]Y^{-1}(i,k), \tag{5.18}$$
$$\hat{B}_{ik} = [Y^{-1}(i,k) - X(i,k)]^{-1}F(i,k), \tag{5.19}$$
$$\hat{C}_{ik} = L(i,k)Y^{-1}(i,k). \tag{5.20}$$

Proof. The results can be obtained employing the same technique used in Chapter 4. □

It should be noted that terms $\beta_{1_{ik}}X(i,k)$ and $\beta_{1_{ik}}Y(i,k)$ in (5.10)-(5.17) are not convex constraints, which are difficult to solve. The iterative algorithm proposed in Chapter 3 is therefore used to change this non-convex problem into quasi-convex optimization problems, which can be solved effectively by available mathematical tools.

5.4 Numerical Example

Consider the following example, where the plant parameters are described as follows:

$$A = \begin{bmatrix} -1.7 & 3.8 \\ -1 & 1.8 \end{bmatrix},\ B_1 = \begin{bmatrix} 0.1 \\ 0.1 \end{bmatrix},\ B_2 = \begin{bmatrix} 5 \\ 2.01 \end{bmatrix},$$
$$C_1 = \begin{bmatrix} 1 & 0 \end{bmatrix},\ C_2 = \begin{bmatrix} 10.1 & 4.5 \end{bmatrix}, D_1 = 0.1,\ D_2 = 0,$$
$$H_1 = \begin{bmatrix} 0.01 \\ 0 \end{bmatrix},\ H_2 = H_3 = 0.01,\ E_1 = \begin{bmatrix} 1 & 0 \end{bmatrix},\ E_2 = -1,\ E_3 = -1.$$

In our simulation, we assume the sampling period is 0.01 for both sensor and actuation channels, that is, $h^a = h^s = 0.01$, and $n^s = n^a = 0$ which means no data packet dropout happens in the communication channel. From this, it is not hard to see that the longer the sampling period is or the more data packets lost, the smaller the time-delay the communication channel can tolerate. Delay free attenuation constant γ_{d_f} is set to be 1, while constants ε_1, ε_2, ε_3, ε_4, ε_5, ε_6, ε_7, ε_8, and ε_9 are set be equivalent to 1.

In the following simulation, we assume $F(t) = \sin t$ and it can be seen that $\|F(t)\| \leq 1$.

The random time-delays exist in $\mathscr{S} = \{1,2\}$ and $\mathscr{W} = \{1,2\}$, and their transition rate matrices are given by:

$$\Lambda = \begin{bmatrix} -3 & 3 \\ 2 & -2 \end{bmatrix},\ \Pi = \begin{bmatrix} -1 & 1 \\ 2 & -2 \end{bmatrix}.$$

Furthermore, we assume that the sensor-to-controller communication delays for two Markovian modes are $|\tau_1^s| < 0.02$, $|\tau_2^s| < 0.015$, while the controller-to-actuator delays are $|\tau_1^a| < 0.02$, and $|\tau_2^a| < 0.015$, and therefore by (2.6) and (2.7) we can have $\hbar_{11} = 0.06$, $\hbar_{12} = 0.055$, $\hbar_{21} = 0.055$, and $\hbar_{22} = 0.05$. By applying Theorem 5.1 and the algorithm in the previous section, we get the following controller gains by the calculation of (5.18)-(5.20):

$$\hat{A}_{11} = \begin{bmatrix} -21.3803 & 38.0585 \\ -11.0071 & 14.3738 \end{bmatrix}, \hat{B}_{11} = \begin{bmatrix} 0.1753 \\ 0.2856 \end{bmatrix}, \hat{C}_{11} = \begin{bmatrix} -3.5890 & 7.0295 \end{bmatrix},$$
$$\hat{A}_{12} = \begin{bmatrix} -25.8097 & 44.6593 \\ -15.0778 & 14.7862 \end{bmatrix}, \hat{B}_{12} = \begin{bmatrix} 0.3307 \\ 0.6163 \end{bmatrix}, \hat{C}_{12} = \begin{bmatrix} -4.1940 & 8.5718 \end{bmatrix},$$
$$\hat{A}_{21} = \begin{bmatrix} -17.5245 & 33.8142 \\ -9.0817 & 13.1192 \end{bmatrix}, \hat{B}_{21} = \begin{bmatrix} 0.0136 \\ 0.1733 \end{bmatrix}, \hat{C}_{21} = \begin{bmatrix} -3.1399 & 6.0225 \end{bmatrix},$$
$$\hat{A}_{22} = \begin{bmatrix} -24.1805 & 45.3803 \\ -12.9956 & 16.7991 \end{bmatrix}, \hat{B}_{22} = \begin{bmatrix} 0.1464 \\ 0.3631 \end{bmatrix}, \hat{C}_{22} = \begin{bmatrix} -4.2063 & 8.4647 \end{bmatrix}.$$

The ratio of the regulated output energy to the disturbance input noise is depicted in Figure 5.2. In our simulation, we use a uniform distributed random disturbance input signal $w(t)$ with maximum value 3. The mode transition of the controller during the simulation is depicted in Figure 5.3 with initial mode 1. Mode 1 represents controller $\mathscr{G}_{11}$, mode 2 for $\mathscr{G}_{12}$, mode 3 for $\mathscr{G}_{21}$, and mode 4 for $\mathscr{G}_{22}$. It can be seen that the ratio tends to a constant value of about 8.75×10^{-5}, which means the attenuation level equals $\sqrt{8.75 \times 10^{-5}} \approx 9.35 \times 10^{-3}$, less than the prescribed level $\gamma = \sqrt{\gamma_{d_f} + \max(\hbar_{ik})} = \sqrt{1+0.06} \approx 1.03$. Furthermore, under the same assumptions on the sampling period h^a and h^s, we choose $n^s = 2$ and $n^a = 1$ to model the data dropouts in the communication channel. In this case, sensor-to-controller delays and controller-to-actuator delays are under the same bound as in the previous case, and hence we can have $\hbar_{11} = 0.09$, $\hbar_{12} = 0.085$, $\hbar_{21} = 0.085$, and $\hbar_{22} = 0.08$. The controller parameters are obtained as:

$$\hat{A}_{11} = \begin{bmatrix} -23.661 & 40.98 \\ -12.868 & 15.098 \end{bmatrix}, \hat{B}_{11} = \begin{bmatrix} 0.22293 \\ 0.3996 \end{bmatrix}, \hat{C}_{11} = \begin{bmatrix} -3.9494 & 7.6575 \end{bmatrix},$$
$$\hat{A}_{12} = \begin{bmatrix} -39.8 & 71.611 \\ -21.756 & 24.545 \end{bmatrix}, \hat{B}_{12} = \begin{bmatrix} 0.34896 \\ 0.75303 \end{bmatrix}, \hat{C}_{12} = \begin{bmatrix} -6.976 & 14.013 \end{bmatrix},$$
$$\hat{A}_{21} = \begin{bmatrix} -17.824 & 32.203 \\ -9.4566 & 12.368 \end{bmatrix}, \hat{B}_{21} = \begin{bmatrix} 0.098429 \\ 0.23225 \end{bmatrix}, \hat{C}_{21} = \begin{bmatrix} -3.0285 & 5.7766 \end{bmatrix},$$
$$\hat{A}_{22} = \begin{bmatrix} -26.6 & 48.909 \\ -14.265 & 17.948 \end{bmatrix}, \hat{B}_{22} = \begin{bmatrix} 0.14056 \\ 0.41352 \end{bmatrix}, \hat{C}_{22} = \begin{bmatrix} -4.5944 & 9.166 \end{bmatrix}.$$

The relative simulation result of the ratio of the regulated output energy to the disturbance input noise is shown in Figure 5.4 where the attenuation level approximately equals to $\sqrt{1.1 \times 10^{-4}} \approx 0.01$, still less than the prescribed level $\gamma = \sqrt{\gamma_{d_f} + \max(\hbar_{ik})} = \sqrt{1+0.09} \approx 1.044$, but worse than the case where there are

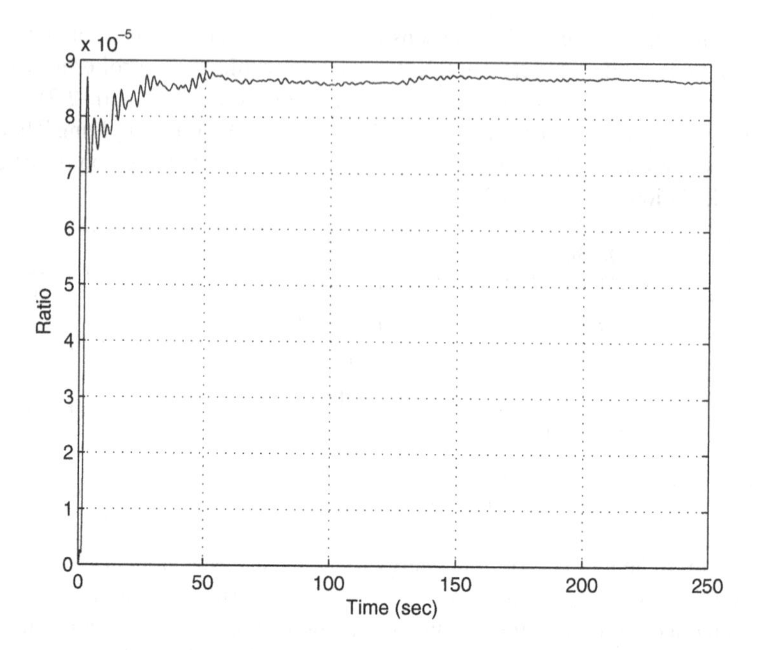

Fig. 5.2 The ratio of the regulated output energy to the disturbance input noise without data dropouts

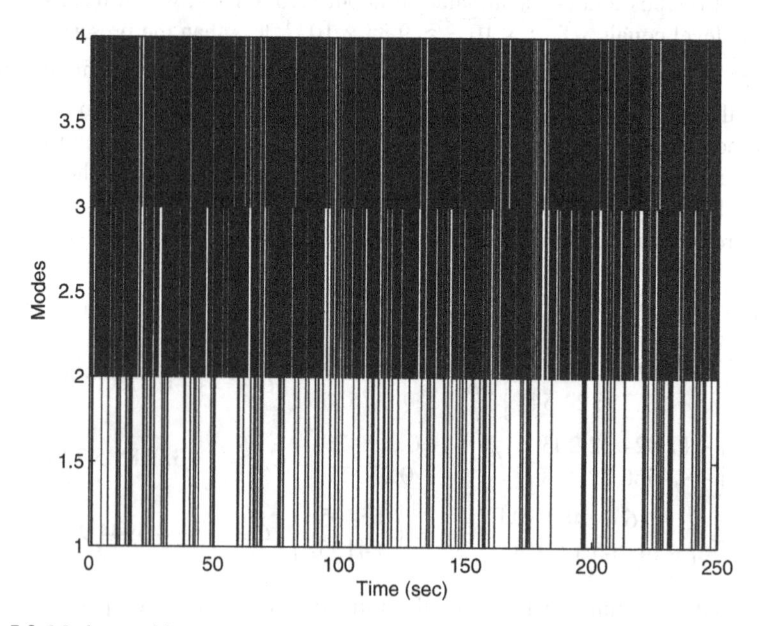

Fig. 5.3 Mode transitions

no data dropouts. The compared result shows that the data dropouts in the communication channel reduce the system performance.

In conclusion, the designed controller meets the performance requirements.

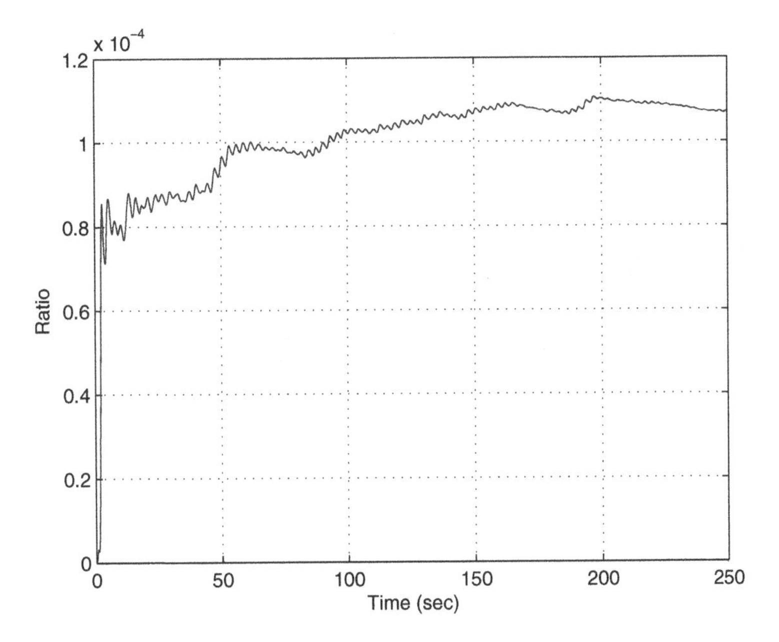

Fig. 5.4 The ratio of the regulated output energy to the disturbance input noise with data dropouts

5.5 Conclusion

In this chapter, a technique of designing a delay-dependant dynamic output feedback controller with robust disturbance attenuation and stability for an uncertain NCS with random communication network-induced delays and data packet dropouts has been proposed. The main contribution of this work is that both the sensor-to-controller and controller-to-actuator delays/dropouts have been taken into account. Furthermore, these delays are regarded as input delays and are dealt with in the scope of disturbance attenuation. The Lyapunov–Razumikhin method has been employed to derive such a controller for this class of systems. Sufficient conditions for the existence of such a controller for this class of NCSs are derived. We finally use a numerical example to demonstrate the effectiveness of this methodology in the last section.

Chapter 6
Robust Filter Design for Uncertain Networked Control Systems

6.1 Introduction

Knowing the system state is necessary to solve many control theory problems. In most practical cases, however, the physical state of the system cannot be determined by direct observation. Instead, indirect effects of the internal state are observed by way of the system outputs. In this sense, a filter is a system that estimates the values of internal systems variables that are not measured from the available outputs [86, 115]. More precisely, the aim of a filter design is that the induced operator norm of the mapping from the noise to the filter error is kept within a prescribed bound.

However, the filter design problem becomes complicated once a communication network is introduced in the system setup. As mentioned in previous chapters, the network-induced effects introduce levels of complexity to the analysis and design of the control systems, because for the filter design problem for NCSs, the main concern is that the state variation depends not only on the current but also the history of exterior disturbance input.

We attempt to solve this problem is this chapter. Our object is to design a robust filter such that the $\mathscr{L}_2$ gain from an exogenous input to an filter error output is less than or equal to a prescribed value, irrespective of the uncertainties and network-induced effects, i.e., network-induced delays and packet dropouts in the sensor-filter channels. Based on the Lyapunov–Razumikhin method, the existence of a delay-dependent filter is given in terms of the solvability of BMIs.

6.2 System Description and Problem Formulation

In this chapter, we assume that $u(t) = 0$ without loss of generality. Assume that the uncertain linear continuous state-space model of the plant dynamics is described by the following equations:

D. Huang and S.K. Nguang: Robust Ctrl. for Uncertain Networked Ctrl. Sys., LNCIS 386, pp. 65–72.
springerlink.com

$$\begin{cases} \dot{x}(t) = (A + \Delta A)x(t) + (B_1 + \Delta B_1)w(t) \\ z(t) = (C_1 + \Delta C_1)x(t) \\ y(t) = (C_2 + \Delta C_2)x(t) + (D_2 + \Delta D_2)w(t) \end{cases} \tag{6.1}$$

where $x(t) \in \mathbb{R}^n$ is the state vector, $w(t) \in \mathbb{R}^p$ is the exogenous disturbance input and/or measurement noise, $y(t) \in \mathbb{R}^l$ and $z(t) \in \mathbb{R}^s$ denote the measurement and regulated output respectively. Matrices A, B_1, C_1, C_2, and D_2 are of appropriate dimensions.

Matrices ΔA, ΔB_1, ΔC_1, ΔC_2, and ΔD_2 characterize the uncertainties in the system and satisfy the following assumption:

Assumption 6.1.

$$\begin{aligned} \begin{bmatrix} \Delta A & \Delta B_1 \end{bmatrix} &= H_1 F(t) \begin{bmatrix} E_1 & E_2 \end{bmatrix}, \\ \Delta C_1 &= H_2 F(t) E_1, \\ \begin{bmatrix} \Delta C_2 & \Delta D_2 \end{bmatrix} &= H_3 F(t) \begin{bmatrix} E_1 & E_2 \end{bmatrix}, \end{aligned}$$

where H_1, H_2, H_3, E_1, and E_2 are known real constant matrices of appropriate dimensions, and $F(t)$ is an unknown matrix function with Lebesgue-measurable elements and satisfies $F(t)^T F(t) \leq I$, in which I is the identity matrix of appropriate dimension.

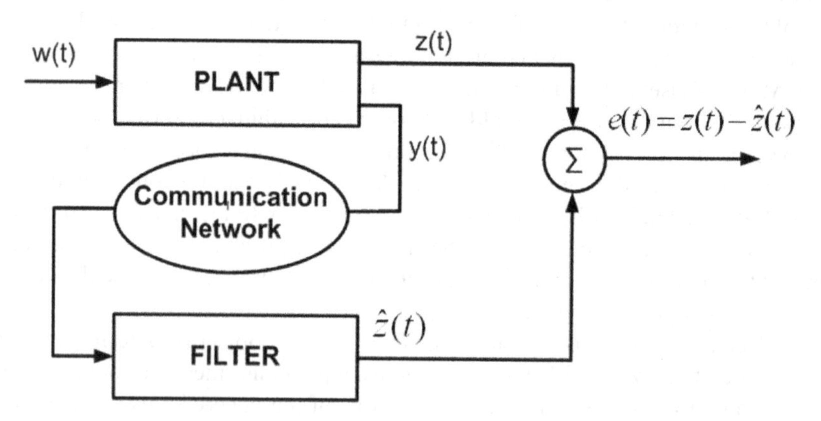

Fig. 6.1 Block diagram of an uncertain NCS with a robust filter

In this chapter, we consider an NCS with a robust filter of which the setup is depicted in Figure 6.1. Following the same lines in Chapter 2 with regard to the modeling of NCSs, a robust filter can be constructed as follows:

$$\mathscr{F}_{\eta(t)} : \begin{cases} \dot{\hat{x}}(t) = \hat{A}(\eta(t))\hat{x}(t) + \hat{B}(\eta(t))y(t - \tau((\eta(t)),t)), \\ \hat{z}(t) = \hat{C}(\eta(t))\hat{x}(t), \end{cases} \tag{6.2}$$

Matrices $\hat{A}(\iota)$, $\hat{B}(\iota)$, $\hat{C}(\iota)$ are the filter's parameters.

Therefore, with regard to (6.1) and (6.2), they can be written in the following concise form:

$$\dot{\tilde{x}}(t) = \mathscr{A}(\eta(t))\tilde{x}(t) + \mathscr{B}(\eta(t))\tilde{x}(t-\tau(\eta(t),t)) + \mathscr{C}(\eta(t))\omega(t), \tag{6.3}$$

$$e(t) = z(t) - \hat{z}(t) = \mathscr{D}(\eta(t))\tilde{x}(t), \tag{6.4}$$

where $\tilde{x}(t) = [x^T(t) \;\; \hat{x}^T(t)]^T$, $w(t-\tau(\eta(t),t)) = v(t)$, $\omega(t) = [w^T(t) \;\; v^T(t)]^T$, and

$$\mathscr{A}(\eta(t)) = \begin{bmatrix} A+\Delta A & 0 \\ 0 & \hat{A}(\eta(t)) \end{bmatrix},$$

$$\mathscr{B}(\eta(t)) = \begin{bmatrix} 0 & 0 \\ \hat{B}(\eta(t))(C_2+\Delta C_2) & 0 \end{bmatrix},$$

$$\mathscr{C}(\eta(t)) = \begin{bmatrix} B_1+\Delta B_1 & 0 \\ 0 & \hat{B}(\eta(t))(D_2+\Delta D_2) \end{bmatrix},$$

$$\mathscr{D}(\eta(t)) = \begin{bmatrix} C_1+\Delta C_1 & -\hat{C}(\eta(t)) \end{bmatrix}.$$

The problem addressed in this chapter is to design a filter of the form (6.2) such that the induced operator norm of the mapping from the noise $w(t)$ to the filter error $e(t) = z(t) - \hat{z}(t)$ is kept within a prescribed bound. It should be noted that with the insertion of a communication network, the analysis has to include the history of the noise, which introduces levels of complexity.

We now cite the the Razumikhin-type theorem established in [105] for the stochastic systems with Markovian jumps to present our problem formulation.

Let $C^{2,1}(\mathbb{R}^n \times \mathscr{S} \times [-\tau,\infty);\mathbb{R}_+)$ denote the family of all nonnegative functions $V(x(t),\eta(t),t)$ on $\mathbb{R}^n \times \mathscr{S} \times [-\tau,\infty)$, which are continuously twice differentiable in x and once differentiable in t. Also, let $\tau > 0$ and $C([-\tau,0];\mathbb{R}^n)$ denote the family of continuous function φ from $[-\tau,0]$ to $\mathbb{R}^n$ with the norm $\| \varphi \| = \sup_{-\tau\leq\theta\leq 0} | \varphi(\theta) |$. Let $(\Omega,\mathscr{F},\{\mathscr{F}_t\}_{t\geq 0},P)$ be a complete probability space with a filtration $\{\mathscr{F}_t\}_{t\geq 0}$ satisfying the usual conditions. Denote by $L^2_{\mathscr{F}_t}([-\tau,0];\mathbb{R}^n)$ the family of all $\{\mathscr{F}_t\}$-measurable $C([-\tau,0];\mathbb{R}^n)$-valued random variables $\phi = \{\phi(\theta) : -\tau \leq \theta \leq 0\}$ such that $\sup_{-\tau\leq\theta\leq 0} \mathbf{E} | \phi(\theta) |^2 < \infty$.

Definition 6.1. Let $\zeta, \alpha_1, \alpha_2$ be all positive numbers and $\delta > 1$. Assume that there exists a function $V \in C^{2,1}(\mathbb{R}^n \times \mathscr{S} \times [-\chi,\infty);\mathbb{R}_+)$ such that

$$\alpha_1 \|x(t)\|^2 \leq V(x(t),\eta(t),t) \leq \alpha_2 \|x(t)\|^2, \tag{6.5}$$

for all $(x(t),\eta(t),t) \in \mathbb{R}^n \times \mathscr{S} \times \mathscr{W} \times [-\chi,\infty)$, and also for system (6.1), if its zero state response $(x(\phi) = 0,\ \omega(\phi) = 0,\ -\chi \leq \phi \leq 0)$ with filter (6.2) satisfies,

$$\mathbf{E}\Big[\int_0^{T_f} \Big(z(t)-\hat{z}(t)\Big)^T \Big(z(t)-\hat{z}(t)\Big) dt\Big]$$
$$\leq \gamma^2 \mathbf{E}\Big[\int_0^{T_f} \sup_{-\chi\leq\phi\leq 0} \omega^T(t+\phi)\omega(t+\phi) dt\Big], \tag{6.6}$$

for any nonzero $\omega(t) \in \mathscr{L}_2[0, T_f]$ and $T_f \geq 0$, provided $x = \{x(\xi) : t - 2\chi \leq \xi \leq t\} \in L^2_{\mathscr{F}_t}([-2\chi, 0]; \mathbb{R}^n)$ satisfying:

$$\mathbf{E}\Big[\min_{\eta(t)\in\mathscr{S}\in\mathscr{W}} V(x(\xi),\eta(\xi),\xi)\Big] < \delta\mathbf{E}\Big[\max_{\eta(t)\in\mathscr{S}\in\mathscr{W}} V(x(t),\eta(t),t)\Big], \quad (6.7)$$

for all $t - 2\chi \leq \xi \leq t$. Then the filter (6.2) is said to satisfy the prescribed $\mathscr{H}_\infty$ performance level $\gamma > 0$.

From here, we use (*) as an ellipsis for terms that are induced by symmetry in the symmetric block matrices. For notation simplicity, we will denote $\mathscr{A}(\eta(t)) = \mathscr{A}_i$ when $\eta(t) = i \in \mathscr{S}$, and wherever appropriate.

6.3 Main Result

In this chapter, we assume that τ_k^s is bounded to $\tau^*(i)$, which is known positive constants. The following theorem provides sufficient conditions for the existence of a robust filter of the form (6.2) for the system (6.1) that satisfy $\mathscr{H}_\infty$ requirement.

Theorem 6.1. *Consider the system (6.1) satisfying Assumption 6.1. For given positive delay free attenuation constant γ_{d_f}, positive constants $\tau^*(i)$, ε_1, ε_2, ε_3, ε_4, ε_5, ε_6, and ε_7, if there exist symmetric positive matrices $X(i)$, $Y(i)$, R_{1_i}, R_{2_i}, R_{3_i}, and R_{4_i}, and matrices $F(i)$, and positive scalars β_{1_i}, β_{2_i}, such that the following inequalities (6.8)-(6.14) hold where $i \in \mathscr{S}$:*

$$\begin{bmatrix} Y(i) & I \\ I & X(i) \end{bmatrix} > 0, \quad (6.8)$$

$$\Psi_i < 0, \quad (6.9)$$

$$\Phi_i < 0, \quad (6.10)$$

$$\begin{bmatrix} R_{4_i} & (*)^T \\ \Lambda_i^T & \mathscr{Q}_i \end{bmatrix} > 0, \quad (6.11)$$

$$\begin{bmatrix} -R_{1_i} & (*)^T & (*)^T & (*)^T \\ 0 & -I & (*)^T & (*)^T \\ 0 & -Y(i) & -R_{2_i} & (*)^T \\ 0 & 0 & 0 & -R_{3_i} \end{bmatrix} < 0, \quad (6.12)$$

$$\begin{bmatrix} -\beta_{2_i}Y(i) & (*)^T & (*)^T & (*)^T & (*)^T & (*)^T \\ -\beta_{2_i}I & -\beta_{2_i}X(i) & (*)^T & (*)^T & (*)^T & (*)^T \\ 0 & Y(i)C_2^T F^T(i) & -Y(i) & (*)^T & (*)^T & (*)^T \\ 0 & 0 & -I & -X(i) & (*)^T & (*)^T \\ 0 & \varepsilon_5 H_3^T F^T(i) & 0 & 0 & -\varepsilon_5 I & (*)^T \\ 0 & 0 & E_1 Y(i) & 0 & 0 & -\varepsilon_5 I \end{bmatrix} < 0, \quad (6.13)$$

$$\begin{bmatrix} -Y(i) & (*)^T & (*)^T & (*)^T & (*)^T & (*)^T & (*)^T & (*)^T \\ -I & -X(i) & (*)^T & (*)^T & (*)^T & (*)^T & (*)^T & (*)^T \\ B_1^T & B_1^T X(i) & -I & (*)^T & (*)^T & (*)^T & (*)^T & (*)^T \\ 0 & D_2^T F^T(i) & 0 & -I & (*)^T & (*)^T & (*)^T & (*)^T \\ \varepsilon_6 H_1^T & \varepsilon_6 H_1^T X(i) & 0 & 0 & -\varepsilon_6 I & (*)^T & (*)^T & (*)^T \\ 0 & 0 & E_2 & 0 & 0 & -\varepsilon_6 I & (*)^T & (*)^T \\ 0 & \varepsilon_7 H_3^T F^T(i) & 0 & 0 & 0 & 0 & -\varepsilon_7 I & (*)^T \\ 0 & 0 & 0 & E_2 & 0 & 0 & 0 & -\varepsilon_7 I \end{bmatrix} < 0, \quad (6.14)$$

where

$$\Psi_i = \left[\begin{matrix} \begin{pmatrix} AY(i)+Y(i)A^T \\ +(\beta_{1_i}+4\beta_{2_i})\tau^*(i)Y(i) \\ +\lambda_{ii}Y(i) \end{pmatrix} & (*)^T & (*)^T & (*)^T \\ (\beta_{1_i}+4\beta_{2_i})\tau^*(i)I & \begin{pmatrix} X(i)A+A^T X(i) \\ +F(i)C_2+C_2^T F^T(i) \\ +(\beta_{1_i}+4\beta_{2_i})\tau^*(i)X(i) \\ +\sum_{j=1}^{s}\lambda_{ij}X(j) \end{pmatrix} & (*)^T & (*)^T \\ B_1^T & B_1^T X(i) & -\gamma_{d_f} I & (*)^T \\ 0 & D_2^T F^T(i) & 0 & -\gamma_{d_f} I \\ C_1 Y(i) - L(i) & C_1 & 0 & 0 \\ E_1 Y(i) & E_1 & E_2 & 0 \\ \varepsilon_1 H_1^T & \varepsilon_1 H_1^T X(i) & 0 & 0 \\ E_1 Y(i) & E_1 & 0 & E_2 \\ 0 & \varepsilon_2 H_3^T F^T(i) & 0 & 0 \\ E_1 Y(i) & E_1 & 0 & 0 \\ 0 & 0 & 0 & 0 \\ S^T(i) & 0 & 0 & 0 \end{matrix}\right.$$

$$\left.\begin{matrix} (*)^T & (*)^T & (*)^T & (*)^T & (*)^T & (*)^T & (*)^T & (*)^T \\ (*)^T & (*)^T & (*)^T & (*)^T & (*)^T & (*)^T & (*)^T & (*)^T \\ (*)^T & (*)^T & (*)^T & (*)^T & (*)^T & (*)^T & (*)^T & (*)^T \\ (*)^T & (*)^T & (*)^T & (*)^T & (*)^T & (*)^T & (*)^T & (*)^T \\ -I & (*)^T & (*)^T & (*)^T & (*)^T & (*)^T & (*)^T & (*)^T \\ 0 & -\varepsilon_1 I (*)^T & (*)^T & (*)^T & (*)^T & (*)^T & (*)^T & \\ 0 & 0 & -\varepsilon_1 I & (*)^T & (*)^T & (*)^T & (*)^T & (*)^T \\ 0 & 0 & 0 & -\varepsilon_2 I & (*)^T & (*)^T & (*)^T & (*)^T \\ 0 & 0 & 0 & 0 & -\varepsilon_2 I & (*)^T & (*)^T & (*)^T \\ 0 & 0 & 0 & 0 & 0 & -\varepsilon_3 I & (*)^T & (*)^T \\ \varepsilon_3 H_2^T & 0 & 0 & 0 & 0 & 0 & -\varepsilon_3 I & (*)^T \\ 0 & 0 & 0 & 0 & 0 & 0 & 0 & -\mathcal{Q}_i \end{matrix}\right],$$

$$\Phi_i = \begin{bmatrix} -\beta_{1_i} Y(i) + R_{1_i} & (*)^T & (*)^T & (*)^T & (*)^T & (*)^T & (*)^T \\ -\beta_{1_i} I & -\beta_{1_i} X(i) & (*)^T & (*)^T & (*)^T & (*)^T & (*)^T \\ Y(i)A^T & -A - Y(i)C_2^T F^T(i) - \lambda_{ii} I & -Y(i) + R_{2_i} & (*)^T & (*)^T & (*)^T & (*)^T \\ A^T & A^T X(i) & -I & -X(i) + R_{3_i} & (*)^T & (*)^T & (*)^T \\ \varepsilon_3 H_1^T & \varepsilon_3 H_1^T X(i) & 0 & 0 & -\varepsilon_4 I & (*)^T & (*)^T \\ 0 & 0 & E_1 Y(i) & E_1 & 0 & -\varepsilon_4 I & (*)^T \\ 0 & R_{4_i} & 0 & 0 & 0 & 0 & -I \end{bmatrix},$$

with

$$S(i) = [\sqrt{\lambda_{i1}} Y(i) \cdots \sqrt{\lambda_{i(i-1)}} Y(i) \sqrt{\lambda_{i(i+1)}} Y(i) \cdots \sqrt{\lambda_{is}} Y(i)],$$
$$\Lambda_i = [\sqrt{\lambda_{i1}} I \cdots \sqrt{\lambda_{i(i-1)}} I \ \sqrt{\lambda_{i(i+1)}} I \cdots \sqrt{\lambda_{is}} I],$$

and

$$\mathscr{Q}_i = diag\{Y(1), \cdots, Y(i-1), Y(i+1), \cdots, Y(s)\},$$

then (6.6) holds for all delays $\tau(i,t)$ satisfying $\tau(i,t) \leq \tau^*(i)$ with $\gamma^2 = \gamma_{d_f} + \max(\tau^*(i))$. Furthermore, the mode dependant robust filter $\mathscr{F}_i$ is of the form (6.2) with

$$\begin{aligned} \hat{A}_i &= [Y^{-1}(i) - X(i)]^{-1}[-A^T - X(i)AY(i) \\ &\quad -F(i)C_2 Y(i) - \sum_{j=1}^{s} \lambda_{ij} Y^{-1}(j) Y(i)] Y^{-1}(i), && (6.15) \\ \hat{B}_i &= [Y^{-1}(i) - X(i)]^{-1} F(i), && (6.16) \\ \hat{C}_i &= L(i) Y^{-1}(i). && (6.17) \end{aligned}$$

Proof. The results can be obtained employing the same technique used in Chapter 4. It is basically an application of Lyapunov stability theorem and Razumikhin-type theorem [105] for stochastic systems with Markovian jumps. □

It should be noted that terms $\beta_{1_i} X(i)$ and $\beta_{1_i} Y(i)$ in (6.9)-(6.14) are not convex constraints, which are difficult to solve. The iterative algorithm proposed in Chapter 3 is therefore used to change this non-convex problem into quasi-convex optimization problems, which can be solved effectively by available mathematical tools.

6.4 Numerical Example

Consider the following numerical example, where the plant parameters are described as follows:

$$A = \begin{bmatrix} -0.1 & -2 \\ -1 & -10 \end{bmatrix}, B_1 = \begin{bmatrix} 0 \\ 0.1 \end{bmatrix}, C_1 = \begin{bmatrix} 1 & 0 \end{bmatrix}, C_2 = \begin{bmatrix} 1 & 0 \end{bmatrix},$$

$$D_2 = 0, H_1 = \begin{bmatrix} 0.01 \\ 0 \end{bmatrix}, H_2 = H_3 = 0.01, E_1 = \begin{bmatrix} 1 & 0 \end{bmatrix}, E_2 = -1.$$

In our simulation, we assume the sampling period is 0.01 for the sensor-to-filter channel, that is, h^s, and $n^s = 0$ which means no data packet dropout happens in the communication channel. Delay free attenuation constant γ_{d_f} is set to be 1, while constants ε_1, ε_2, ε_3, ε_4, ε_5, ε_6, and ε_7 are set be equivalent to 1.

In the following simulation, we assume $F(t) = \sin t$ and it can be seen that $\|F(t)\| \leq 1$.

The random time-delays exist in two modes, i.e., $i \in \mathscr{S} = \{1,2\}$, and their transition rate matrices are given by:

$$\Lambda = \begin{bmatrix} -3 & 3 \\ 2 & -2 \end{bmatrix}.$$

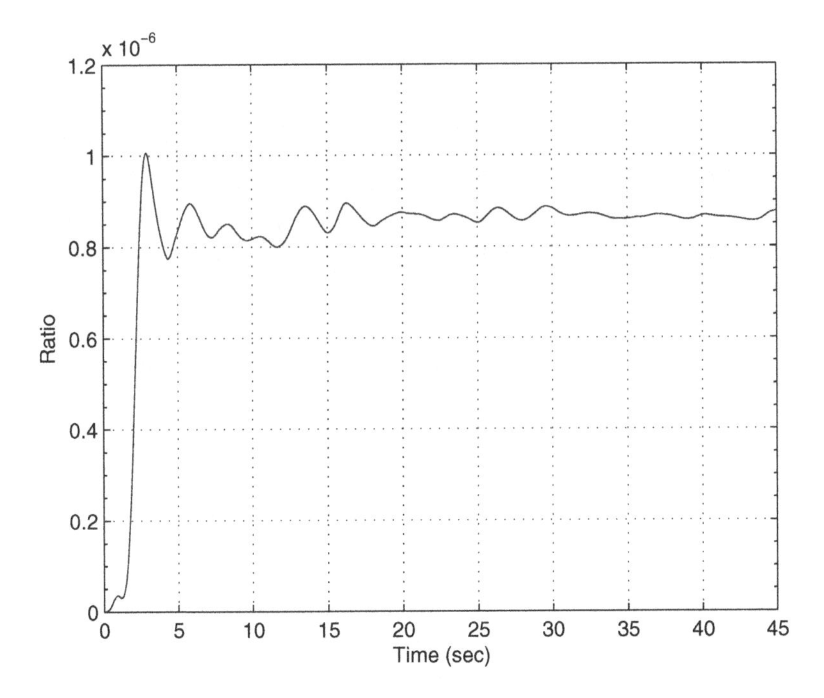

Fig. 6.2 The ratio of the filter error energy to the disturbance noise energy without data dropouts

Furthermore, we assume that the sensor-to-filter communication delays for the two Markovian modes are $\tau^*(1) < 0.02$, $\tau^*(2) < 0.015$. By applying Theorem 6.1

and the algorithm in the previous section, we get the following controller gains by the calculation of (6.15)-(6.17):

$$\hat{A}_1 = \begin{bmatrix} 17.5833 & -3.6094 \\ -6.3456 & 2.0987 \end{bmatrix}, \hat{B}_1 = \begin{bmatrix} 1.8932 \\ -0.2438 \end{bmatrix}, \hat{C}_1 = \begin{bmatrix} -13.6735 & 1.2239 \end{bmatrix},$$

$$\hat{A}_2 = \begin{bmatrix} 19.3651 & -3.2232 \\ -5.9803 & 3.1127 \end{bmatrix}, \hat{B}_2 = \begin{bmatrix} 2.0985 \\ -0.3547 \end{bmatrix}, \hat{C}_2 = \begin{bmatrix} -12.1142 & 1.9092 \end{bmatrix}.$$

The ratio of the filter error energy to the disturbance input noise is depicted in Figure 6.2. In our simulation, we use a uniform distributed random disturbance input signal $w(t)$ with maximum value 2. The mode transition of the filter during the simulation is depicted in Figure 6.3 with initial mode 1. Mode 1 represents filter $\mathscr{F}_1$, and mode 2 for $\mathscr{F}_2$. It can be seen that the ratio tends to a constant value of about 0.85×10^{-7}, which means the attenuation level equals to $\sqrt{0.85 \times 10^{-7}} \approx 9.2 \times 10^{-4}$, less than the prescribed level $\gamma = \sqrt{\gamma_{d_f} + \max(\tau^*(i))} = \sqrt{1+0.02} \approx 1.01$.

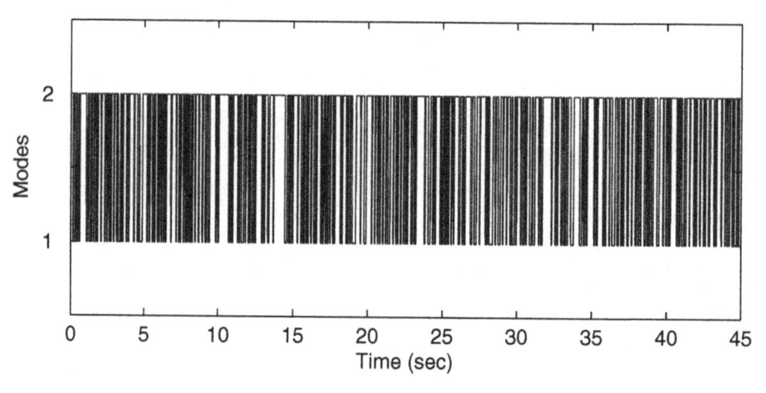

Fig. 6.3 Mode transitions

In conclusion, the designed controller meets the performance requirements.

6.5 Conclusion

In this chapter, a technique of designing a delay-dependant dynamic output feedback filter for an uncertain NCS with random communication network-induced delays and data packet dropouts has been proposed. The main contribution of this work is that network-induced effects are regarded as input delays and are dealt with in the scope of disturbance attenuation. The Lyapunov–Razumikhin method has been employed to derive such a robust filter for this class of systems. Sufficient conditions for the existence of such a filter for this class of NCSs are derived.

Chapter 7
Robust Fault Estimator Design for Uncertain Networked Control Systems

In order to avoid production deteriorations or damages, system faults have to be identified and decisions that stop the propagation of their effects have to be made. This gives the rise to the research on fault detection and isolation (FDI) and in recent years, the problem has attracted lots of attention from researchers. Among them, the model-based approach is the common approach, see survey papers [88, 89, 90]. The prime importance [116, 117] in designing a model-based fault-detection system is the increasing robustness of residual to unknown inputs and modeling errors and enhancing the sensitivity to faults. Two approaches are mainly applied in FDI to address these two issues. One is to use the $\mathscr{H}_\infty$ norm of transfer function matrix from fault to residual signal as a measure to estimate the sensitivity to the faults [119, 120]. Another method is to adopt the $\mathscr{H}_\infty$-filtering formulation to make the the error between residual and fault as small as possible [118, 121]. Furthermore, the existence of time-delays is commonly encountered in dynamic systems and has to be dealt with in the realm of FDI. Some results have been obtained to address this issue, see [91]-[95]. However, these results are mostly obtained for systems with state delays.

Firstly, this chapter applies the disturbance attenuation notation for systems with input delays used in previous chapter. In light of such formulation, this chapter proposes a robust fault estimator that ensures the fault estimation error is less than prescribed performance level, irrespective of the uncertainties and network-induced effects, i.e., network-induced delays and packet dropouts in communication channels, which are to be modeled by the Markov processes. Based on the Lyapunov–Razumikhin method, the existence of a delay-dependent fault estimator is given in terms of the solvability of BMIs. An iterative algorithm is proposed to change this non-convex problem into quasi-convex optimization problems, which can be solved effectively by available mathematical tools.

D. Huang and S.K. Nguang: Robust Ctrl. for Uncertain Networked Ctrl. Sys., LNCIS 386, pp. 73–83.
springerlink.com

7.1 Problem Formulation and Preliminaries

Assume that the uncertain linear continuous state-space model of the plant dynamics is described by the following equations:

$$\begin{cases} \dot{x}(t) = (A + \Delta A)x(t) + Bw(t) + Gf(t) \\ y(t) = (C + \Delta C)x(t) + Dw(t) + Jf(t) \end{cases} \tag{7.1}$$

where $x(t) \in \mathbb{R}^n$ is the state vector, $w(t) \in \mathbb{R}^p$ and $f(t) \in \mathbb{R}^q$ are, respectively, exogenous disturbances and faults which belong to $\mathscr{L}_2[0,\infty)$, $y(t) \in \mathbb{R}^l$ denotes the measurement output. Matrices A, B, C, D, G, and J are of appropriate dimensions.

Matrices ΔA and ΔC characterize the uncertainties in the system and satisfy the following assumption:

Assumption 7.1.

$$\begin{bmatrix} \Delta A \\ \Delta C \end{bmatrix} = \begin{bmatrix} H_1 \\ H_2 \end{bmatrix} F(t)E,$$

where H_1, H_2, and E are known real constant matrices of appropriate dimensions, and $F(t)$ is an unknown matrix function with Lebesgue-measurable elements and satisfies $F(t)^T F(t) \leq I$, in which I is the identity matrix of appropriate dimension.

Following the same lines in Chapter 2 with regard to the modeling of NCS, a fault estimation filter is therefore constructed as follows:

$$\mathscr{G} : \begin{cases} \dot{\hat{x}}(t) = \hat{A}(\eta(t))\hat{x}(t) + \hat{B}(\eta(t))y(t - \tau(t)), \\ r_s(t) = \hat{C}(\eta(t))\hat{x}(t) + \hat{D}(\eta(t))y(t - \tau(t)), \end{cases} \tag{7.2}$$

where $\hat{A}(\eta(t))$, $\hat{B}(\eta(t))$, $\hat{C}(\eta(t))$, and $\hat{D}(\eta(t))$ are mode-dependent controller parameters to be designed.

The overall system setup is depicted in Figure 7.1.

Therefore, the state-space form of the system model (7.1) with the fault estimator (7.2) is given by:

$$\begin{bmatrix} \dot{x}(t) \\ \dot{\hat{x}}(t) \end{bmatrix} = \begin{bmatrix} A + \Delta A & 0 \\ 0 & \hat{A}(\eta(t)) \end{bmatrix} \begin{bmatrix} x(t) \\ \hat{x}(t) \end{bmatrix} + \begin{bmatrix} 0 & 0 \\ \hat{B}(\eta(t))(C + \Delta C) & 0 \end{bmatrix} \begin{bmatrix} x(t - \tau(\eta(t),t)) \\ \hat{x}(t - \tau(\eta(t),t)) \end{bmatrix}$$

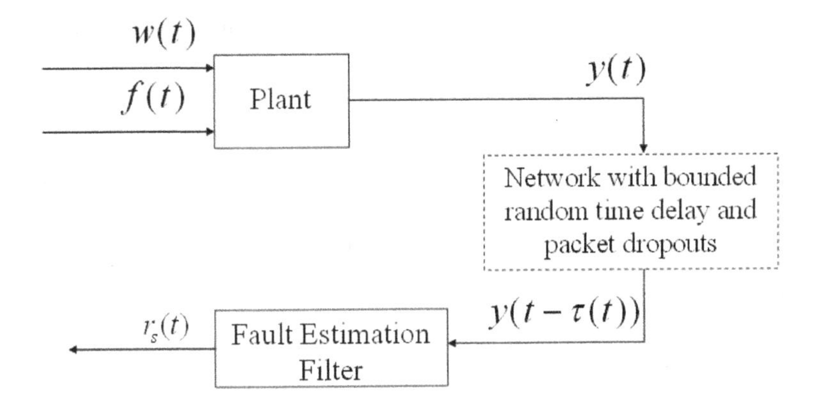

Fig. 7.1 Fault estimation filter for NCS

$$
+\begin{bmatrix} B & G & 0 & 0 \\ 0 & 0 & \hat{B}(\eta(t))D & \hat{B}(\eta(t))J \end{bmatrix} \begin{bmatrix} w(t) \\ f(t) \\ w(t-\tau(\eta(t),t)) \\ f(t-\tau(\eta(t),t)) \end{bmatrix}, \tag{7.3}
$$

$$
e(t) = \begin{bmatrix} 0 & \hat{C}(\eta(t)) \end{bmatrix} \begin{bmatrix} x(t) \\ \hat{x}(t) \end{bmatrix} + \begin{bmatrix} \hat{D}(\eta(t))(C+\Delta C) & 0 \end{bmatrix} \begin{bmatrix} x(t-\tau(\eta(t),t)) \\ \hat{x}(t-\tau(\eta(t),t)) \end{bmatrix}
$$

$$
+\begin{bmatrix} 0 & -I & \hat{D}(\eta(t))D & \hat{D}(\eta(t))J \end{bmatrix} \begin{bmatrix} w(t) \\ f(t) \\ w(t-\tau(\eta(t),t)) \\ f(t-\tau(\eta(t),t)) \end{bmatrix}, \tag{7.4}
$$

where $e(t) = r_s(t) - f(t)$ is the fault estimation error.

Define $\tilde{x}(t) = [x^T(t) \;\; \hat{x}^T(t)]^T$ and $\omega(t) = [w^T(t) \;\; f^T(t) \;\; w^T(t-\tau(\eta(t),t)) \;\; f^T(t-\tau(\eta(t),t))]^T$, (7.3) and (7.4) can be written in the following compact form :

$$
\dot{\tilde{x}}(t) = \mathscr{A}(\eta(t))\tilde{x}(t) + \mathscr{B}(\eta(t))\tilde{x}(t-\tau(\eta(t),t)) + \mathscr{C}(\eta(t))\omega(t) \tag{7.5}
$$

$$
e(t) = \mathscr{D}_1(\eta(t))\tilde{x}(t) + \mathscr{D}_2(\eta(t))\tilde{x}(t-\tau(\eta(t),t)) + \mathscr{D}_3(\eta(t))\omega(t) \tag{7.6}
$$

where

$$
\mathscr{A}(\eta(t)) = \begin{bmatrix} A+\Delta A & 0 \\ 0 & \hat{A}(\eta(t)) \end{bmatrix}, \; \mathscr{B}(\eta(t)) = \begin{bmatrix} 0 & 0 \\ \hat{B}(\eta(t))(C+\Delta C) & 0 \end{bmatrix},
$$

$$
\mathscr{C}(\eta(t)) = \begin{bmatrix} B & G & 0 & 0 \\ 0 & 0 & \hat{B}(\eta(t))D & \hat{B}(\eta(t))J \end{bmatrix},
$$

$$
\mathscr{D}_1(\eta(t)) = \begin{bmatrix} 0 & \hat{C}(\eta(t)) \end{bmatrix}, \; \mathscr{D}_2(\eta(t)) = \begin{bmatrix} \hat{D}(\eta(t))(C+\Delta C) & 0 \end{bmatrix},
$$

$$
\mathscr{D}_3(\eta(t)) = \begin{bmatrix} 0 & -I & \hat{D}(\eta(t))D & \hat{D}(\eta(t))J \end{bmatrix}.
$$

The aim of this chapter is to design a fault estimator of the form (7.2) such that the following inequality holds:

For (7.5) and (7.6) with its zero state response ($x(\phi)=0,\ \omega(\phi)=0,\ -\chi\leq\phi\leq 0$),

$$\mathbf{E}\Big[\int_0^{T_f} e^T(t)e(t)dt\Big] \leq \gamma^2\mathbf{E}\Big[\int_0^{T_f} \sup_{-\chi\leq\phi\leq 0} \omega^T(t+\phi)\omega(t+\phi)dt\Big], \tag{7.7}$$

for any nonzero $\omega(t)\in\mathscr{L}_2[0,\ T_f]$ and $T_f\geq 0$, provided $x=\{x(\xi): t-2\chi\leq\xi\leq t\}\in L^2_{\mathscr{F}_t}([-2\chi,0];\mathbb{R}^n)$ satisfying:

$$\mathbf{E}\Big[\min_{\eta(t)\in\mathscr{S}\in\mathscr{W}} V(x(\xi),\eta_1(\xi),\eta_2(\xi),\xi)\Big] < \delta\mathbf{E}\Big[\max_{\eta(t)\in\mathscr{S}\in\mathscr{W}} V(x(t),\eta(t),t)\Big], \tag{7.8}$$

for all $t-2\chi\leq\xi\leq t$, then a fault estimator is designed satisfying a disturbance attenuation level γ.

From here, we use (*) as an ellipsis for terms that are induced by symmetry in the symmetric block matrices. For notation simplicity, we will denote $\mathscr{A}(\eta(t))=\mathscr{A}_i$ when $\eta(t)=i\in\mathscr{S}$, and wherever appropriate.

7.2 Main Result

In this chapter, we assume that τ_k^s is bounded. For each $\eta(t)=i\in\mathscr{S}$ according to (2.6), there is no loss of generality to assume $\tau(i,t)\leq\tau^*(i)$, where $\tau^*(i)$ are known positive constants. The following theorem provides sufficient conditions for the existence of a robust fault estimator for the system (7.5) and (7.6) such that (7.7) is satisfied.

Theorem 7.1. *Consider the system (7.5) and (7.6) satisfying Assumption 7.1. For given positive delay-free attenuation constant γ_{d_f}, positive constants $\tau^*(i)$, ε_1, ε_2, ε_3, and ε_4, if there exist symmetric positive matrices $X(i)$, $Y(i)$, R_{1_i}, R_{2_i}, R_{3_i}, and R_{4_i}, and matrices $F(i)$, $L(i)$, and $\hat{D}(i)$, and positive scalars β_{1_i}, β_{2_i}, such that the following inequalities hold where $i\in\mathscr{S}$:*

$$\begin{bmatrix} Y(i) & I \\ I & X(i) \end{bmatrix} > 0, \tag{7.9}$$

$$\Upsilon(i) < 0, \tag{7.10}$$

$$\begin{bmatrix} -\beta_{1_i}Y(i)+R_{1_i} & (*)^T & (*)^T & (*)^T & (*)^T & (*)^T & (*)^T \\ -\beta_{1_i}I & -\beta_{1_i}X(i) & (*)^T & (*)^T & (*)^T & (*)^T & (*)^T \\ Y(i)A^T & -A-Y(i)C^TF^T(i)-\lambda_{ii}I & -Y(i)+R_{2_i} & (*)^T & (*)^T & (*)^T & (*)^T \\ A^T & A^TX(i) & -I & -X(i)+R_{3_i} & (*)^T & (*)^T & (*)^T \\ \varepsilon_3H_1^T & \varepsilon_3H_1^TX(i) & 0 & 0 & -\varepsilon_3I & (*)^T & (*)^T \\ 0 & 0 & EY(i) & E & 0 & -\varepsilon_3I & (*)^T \\ 0 & R_{4_i} & 0 & 0 & 0 & 0 & -I \end{bmatrix} < 0, \quad (7.11)$$

$$\begin{bmatrix} R_{4_i} & (*)^T \\ \Lambda_i^T & \mathscr{Q}_i \end{bmatrix} > 0, \quad (7.12)$$

$$\begin{bmatrix} -R_{1_i} & (*)^T & (*)^T & (*)^T \\ 0 & -I & (*)^T & (*)^T \\ 0 & -Y(i) & -R_{2_i} & (*)^T \\ 0 & 0 & 0 & -R_{3_i} \end{bmatrix} < 0, \quad (7.13)$$

$$\begin{bmatrix} -\beta_{2_i}Y(i) & (*)^T & (*)^T & (*)^T & (*)^T & (*)^T \\ -\beta_{2_i}I & -\beta_{2_i}X(i) & (*)^T & (*)^T & (*)^T & (*)^T \\ 0 & Y(i)C^TF^T(i) & -Y(i) & (*)^T & (*)^T & (*)^T \\ 0 & C^TF^T(i) & -I & -X(i) & (*)^T & (*)^T \\ 0 & \varepsilon_4H_2^TF^T(i) & 0 & 0 & -\varepsilon_4I & (*)^T \\ 0 & 0 & EY(i) & E & 0 & -\varepsilon_4I \end{bmatrix} < 0, \quad (7.14)$$

$$\begin{bmatrix} -Y(i) & (*)^T & (*)^T & (*)^T & (*)^T & (*)^T \\ -I & -X(i) & (*)^T & (*)^T & (*)^T & (*)^T \\ B^T & B^TX(i) & -I & (*)^T & (*)^T & (*)^T \\ G^T & G^TX(i) & 0 & -I & (*)^T & (*)^T \\ 0 & D^TF^T(i) & 0 & 0 & -I & (*)^T \\ 0 & J^TF^T(i) & 0 & 0 & 0 & -I \end{bmatrix} < 0, \quad (7.15)$$

where

$$
\Upsilon(i)=\left[\begin{array}{cccccccccccccc}
\Xi_1(i) & (*)^T & (*)^T & (*)^T & (*)^T & (*)^T & (*)^T & (*)^T & (*)^T & (*)^T & (*)^T & (*)^T & (*)^T & (*)^T \\
(\beta_{1_i}+4\beta_{2_i})\tau^*(i)I & \Xi_2(i) & (*)^T & (*)^T & (*)^T & (*)^T & (*)^T & (*)^T & (*)^T & (*)^T & (*)^T & (*)^T & (*)^T & (*)^T \\
0 & 0 & -I & (*)^T & (*)^T & (*)^T & (*)^T & (*)^T & (*)^T & (*)^T & (*)^T & (*)^T & (*)^T & (*)^T \\
0 & 0 & 0 & -I & (*)^T & (*)^T & (*)^T & (*)^T & (*)^T & (*)^T & (*)^T & (*)^T & (*)^T & (*)^T \\
B^T & B^TX(i) & 0 & 0 & -\gamma_{d_f}I & (*)^T & (*)^T & (*)^T & (*)^T & (*)^T & (*)^T & (*)^T & (*)^T & (*)^T \\
G^T & G^TX(i) & 0 & 0 & 0 & -\gamma_{d_f}I & (*)^T & (*)^T & (*)^T & (*)^T & (*)^T & (*)^T & (*)^T & (*)^T \\
0 & D^TF(i) & 0 & 0 & 0 & 0 & -\gamma_{d_f}I & (*)^T & (*)^T & (*)^T & (*)^T & (*)^T & (*)^T & (*)^T \\
0 & J^TF(i) & 0 & 0 & 0 & 0 & 0 & -\gamma_{d_f}I & (*)^T & (*)^T & (*)^T & (*)^T & (*)^T & (*)^T \\
L(i) & 0 & \hat{D}(i)C & 0 & 0 & -I & \hat{D}(i)D & \hat{D}(i)J & -I & (*)^T & (*)^T & (*)^T & (*)^T & (*)^T \\
Z^T(i) & 0 & 0 & 0 & 0 & 0 & 0 & 0 & 0 & -\mathcal{Q}(i) & (*)^T & (*)^T & (*)^T & (*)^T \\
\varepsilon_1 H_1^T & \varepsilon_1(H_1^TX^T(i)+H_2^TF^T(i)) & 0 & 0 & 0 & 0 & 0 & 0 & 0 & 0 & -\varepsilon_1 I & (*)^T & (*)^T & (*)^T \\
EY(i) & E & 0 & 0 & 0 & 0 & 0 & 0 & 0 & 0 & 0 & -\varepsilon_1 I & (*)^T & (*)^T \\
0 & 0 & 0 & 0 & 0 & 0 & 0 & 0 & \varepsilon_2 H_2^T\hat{D}^T(i) & 0 & 0 & 0 & -\varepsilon_2 I & (*)^T \\
0 & 0 & E & 0 & 0 & 0 & 0 & 0 & 0 & 0 & 0 & 0 & 0 & -\varepsilon_2 I
\end{array}\right],
$$

and

$$
\begin{aligned}
&Z(i)=[\sqrt{\lambda_{i1}}Y(i)\cdots\sqrt{\lambda_{i(i-1)}}Y(i)\ \sqrt{\lambda_{i(i+1)}}Y(i)\cdots\sqrt{\lambda_{is}}Y(i)],\\
&\Lambda_i=[\sqrt{\lambda_{i1}}I\cdots\sqrt{\lambda_{i(i-1)}}I\ \sqrt{\lambda_{i(i+1)}}I\cdots\sqrt{\lambda_{is}}I],\\
&\mathcal{Q}_i=diag\{Y(1),\cdots,Y(i-1),Y(i+1),\cdots,Y(s)\},\\
&\Xi_1(i)=AY(i)+Y(i)A^T+(\beta_{1_i}+4\beta_{2_i})\tau^*(i)Y(i)+\lambda_{ii}Y(i),\\
&\Xi_2(i)=X(i)A+A^TX(i)+F(i)C+C^TF^T(i)+(\beta_{1_i}+4\beta_{2_i})\tau^*(i)X(i)+\sum_{j=1}^{s}\lambda_{ij}X(j),
\end{aligned}
$$

then (7.7) holds for delay $\tau(i,t)$ satisfying $\tau(i,t)\le\tau^*(i)$ with $\gamma^2=\gamma_{d_f}+\max(\tau^*(i))$ for $i\in\mathcal{S}$. Furthermore, the mode dependant fault estimator $\mathcal{G}_i$ is obtained of the form (7.2) with

$$
\begin{aligned}
\hat{A}_i &= [Y^{-1}(i)-X(i)]^{-1}[-A^T-X(i)AY(i)-F(i)CY(i)\\
&\quad -\sum_{j=1}^{s}\lambda_{ij}Y^{-1}(j)Y(i)]Y^{-1}(i), \qquad (7.16)\\
\hat{B}_i &= [Y^{-1}(i)-X(i)]^{-1}F(i), \qquad (7.17)\\
\hat{C}_i &= L(i)Y^{-1}(i). \qquad (7.18)
\end{aligned}
$$

Proof. The proof process can be derived from Chapter 4. □

It should be noted that terms $Y(i)C^T F^T(i)$, $\beta_{1_i}X(i)$ and $\beta_{1_i}Y(i)$ in (7.10)-(7.15) are not convex constraints, which are difficult to solve. We therefore propose the following algorithm to change this non-convex feasibility problem into quasi-convex optimization problems [122].

Algorithm 7.1. *ILMI algorithm*

Step 1. Find $X(i)$, $Y(i)$, $\hat{D}(i)$, $F(i)$ and $L(i)$ subject to (7.9) and (7.10) with $\tau^*(i) = 0$. Let $n = 1$ and $X_n(i) = X(i)$ and $Y_n(i) = Y(i)$.
Step 2. Solve the following optimization problem for α_n, $\hat{D}(i)$, $F(i)$, and $L(i)$ with the given $\tau^*(i)$ and $X_n(i)$ and $Y_n(i)$ obtained in the previous step:
OP1 : Minimize α_n subject to the following LMI constraints:

$$\text{Left hand-side of (7.10)} - \alpha_n \left[\begin{array}{cc|c} Y_n(i) & I & 0 \\ I & X_n(i) & \\ \hline \multicolumn{2}{c|}{0} & 0 \end{array}\right] < 0 \tag{7.19}$$

and (7.9), (7.11)-(7.15).
Step 3. If $\alpha_n < 0$, $X_n(i)$, $Y_n(i)$ and $\hat{D}(i)$, $F(i)$, and $L(i)$ are a feasible solution to the BMIs and stop.
Step 4. Set $n = n+1$. Solve the following optimization problem for α_n, $X_n(i)$ and $Y_n(i)$ with $\hat{D}(i)$, $F(i)$, and $L(i)$ obtained in the previous step:
OP2 : Minimize α_n subject to LMI constraints (7.19), (7.9), and (7.11)-(7.15).
Step 5. If $\alpha_n < 0$, $X_n(i)$, $Y_n(i)$ and $\hat{D}(i)$, $F(i)$, and $L(i)$ are a feasible solution to the BMIs and stop.
Step 6. Set $n = n+1$. Solve the following optimization problem for $X_n(i)$ and $Y_n(i)$ with α_n, $\hat{D}(i)$, $F(i)$, and $L(i)$ obtained in the previous step:
OP3 : Minimize trace($\begin{bmatrix} Y_n(i) & I \\ I & X_n(i) \end{bmatrix}$) subject to LMI constraints (7.19), (7.9), and (7.11)-(7.15).
Step 7. Let $T_n = \begin{bmatrix} Y_n(i) & I \\ I & X_n(i) \end{bmatrix}$. If $\| T_n - T_{n-1} \| / \| T_n \| < \zeta$, ζ is a prescribed tolerance, go to Step 8. Else, set $n = n+1$, $X_n(i) = X_{n-1}(i)$ and $Y_n(i) = Y_{n-1}(i)$, then go to Step 2.
Step 8. A fault estimator for the system may not be found, stop.

Remark 7.1.

1. In Step 1, the initial data is obtained by assuming that the system has no time-delay.
2. A term $-\alpha_n \left[\begin{array}{cc|c} Y_n(i) & I & 0 \\ I & X_n(i) & \\ \hline \multicolumn{2}{c|}{0} & 0 \end{array}\right]$ is introduced in (7.10) to relax the LMI constraints. It is referred to as $\alpha/2$-stabilizable problem in [123]. If an $\alpha_n < 0$ can be found, the robust fault estimator can be obtained. The rationale behind this concept can also be found in [98].

3. The optimization problem in Step 2 and Step 4 is a generalized eigenvalue minimization problem. These two steps guarantee the progressive reduction of α_n. Step 6 guarantees the convergence of the algorithm.

7.3 Numerical Example

In this section, we consider the following example where the plant is described as follows:

$$\begin{cases} \dot{x}(t) = (A+\Delta A)x(t) + Bw(t) + Gf(t) \\ y(t) = (C+\Delta C)x(t) + Jf(t) \end{cases} \tag{7.20}$$

where:

$$A = \begin{bmatrix} -1.7 & 3.8 \\ -1 & 1.8 \end{bmatrix}, B = \begin{bmatrix} 0 \\ 1 \end{bmatrix} G = \begin{bmatrix} 0 & 0 \\ 1 & 0 \end{bmatrix}, C = \begin{bmatrix} 0 & 1 \end{bmatrix}, J = \begin{bmatrix} 0 & 1 \end{bmatrix},$$

$$H_1 = \begin{bmatrix} 0.01 & 0 \\ 0 & 0.01 \end{bmatrix}, H_2 = \begin{bmatrix} 0.01 & 0 \end{bmatrix}, E = \begin{bmatrix} 1 \\ 0 \end{bmatrix}.$$

In the following simulation, we assume $F(t) = \sin t$ and it can be seen that $\|F(t)\| \leq 1$.

In our simulation, we assume $\tau^*(1) = 0.045$ and $\tau^*(2) = 0.025$. We assume the sampling period is 0.01, that is, $h^s = 0.01$, and $n^s = 0$ which means no data packet dropout happens in the communication channel. From this, it is not hard to see that the longer the sampling period is or the more data packets are lost, the smaller the time-delay the communication channel can tolerate. Delay free attenuation constant γ_{d_f} is set to be 1, while constants ε_1, ε_2, ε_3, ε_4, and ε_5 are set to be 1.

The random time-delays exist in $\mathscr{S} = \{1, 2\}$, and its transition rate matrices are given by:

$$\Lambda = \begin{bmatrix} -1 & 1 \\ 2 & -2 \end{bmatrix}.$$

In this example, the faults are assumed to be low-frequency faults. Hence, weighting functions for the faults are selected as follows:

$$\begin{cases} \dot{x}_f(t) = A_f x_f(t) + B_f \tilde{f}(t) \\ f(t) = C_f x_f(t) \end{cases} \tag{7.21}$$

where $x_f(t) = \begin{bmatrix} x_{f_1}(t) \\ x_{f_2}(t) \end{bmatrix}$, $\tilde{f}(t) = \begin{bmatrix} \tilde{f}_1(t) \\ \tilde{f}_2(t) \end{bmatrix}$ is the all pass fault, $A_f = \begin{bmatrix} -1 & 0 \\ 0 & -1 \end{bmatrix}$, $B_f = \begin{bmatrix} 1 & 0 \\ 0 & 1 \end{bmatrix}$, and $C_f = \begin{bmatrix} 1 & 0 \\ 0 & 1 \end{bmatrix}$.

Augmenting (7.20) with (7.21), we obtain:

$$\begin{cases} \dot{\tilde{x}}(t) = (\tilde{A}+\Delta\tilde{A})\tilde{x}(t) + \tilde{B}w(t) + \tilde{G}\tilde{f}(t) \\ y(t) = (\tilde{C}+\Delta\tilde{C})\tilde{x}(t) \end{cases} \tag{7.22}$$

where $\tilde{x}(t) = \begin{bmatrix} x(t) \\ x_f(t) \end{bmatrix}$ and

$$\tilde{A} = \begin{bmatrix} -1.7 & 3.8 & 0 & 0 \\ -1 & 1.8 & 1 & 0 \\ 0 & 0 & -1 & 0 \\ 0 & 0 & 0 & -1 \end{bmatrix}, \tilde{B} = \begin{bmatrix} 0 \\ 1 \\ 0 \\ 0 \end{bmatrix}, \tilde{G} = \begin{bmatrix} 0 & 0 \\ 1 & 0 \\ 1 & 0 \\ 0 & 1 \end{bmatrix}, \tilde{C} = \begin{bmatrix} 0 & 1 & 0 & 1 \end{bmatrix},$$

$$\tilde{H}_1 = \begin{bmatrix} 0.01 & 0 & 0 & 0 \\ 0 & 0.01 & 0 & 0 \\ 0 & 0 & 0 & 0 \\ 0 & 0 & 0 & 0 \end{bmatrix}, \tilde{H}_2 = \begin{bmatrix} 0.01 & 0 & 0 & 0 \end{bmatrix}, \tilde{E} = \begin{bmatrix} 1 \\ 0 \\ 0 \\ 0 \end{bmatrix}.$$

For the sake of simplicity, $\hat{D}_i$ is assumed to be a zero matrix in this example. By applying Theorem 7.1 and the algorithm in the previous section and calculating (7.16)-(7.18), we get the following fault estimator for $i \in \mathscr{S} = \{1,2\}$:

$$\mathscr{G}_i : \begin{cases} \dot{\hat{x}}(t) = \hat{A}_i \hat{x}(t) + \hat{B}_i y(kh^s),\ \hat{x}(0) = 0, \\ r_s(t) = \hat{C}_i \hat{x}(t) \end{cases} \forall t \in [kh^s + \tau_k^s, (k+1)h^s + \tau_{k+1}^s], \quad (7.23)$$

where:

$$\hat{A}_1 = \begin{bmatrix} -4.0446 & -124.61 & -0.12359 & -120.52 \\ 1.0434 & 106.39 & 1.141 & 105.87 \\ 0.00704 & 43.068 & -0.94306 & 42.702 \\ -0.036613 & -118.59 & -0.097559 & -119.47 \end{bmatrix}, \hat{B}_1 = \begin{bmatrix} -6.9128 \\ -174.11 \\ -0.9051 \\ -167.15 \end{bmatrix},$$

$$\hat{C}_1 = \begin{bmatrix} 0.2549 & 0.6612 & 1.1190 & -0.0000 \\ 0.0864 & -0.0141 & 0.2550 & 0.0000 \end{bmatrix},$$

$$\hat{A}_2 = \begin{bmatrix} -3.8622 & -91.702 & 0.66102 & -88.226 \\ 0.69258 & 108.3 & -0.25119 & 109 \\ -0.15842 & 59.05 & -1.527 & 59.283 \\ 0.12657 & -103.52 & 0.55874 & -104.89 \end{bmatrix}, \hat{B}_2 = \begin{bmatrix} -5.0985 \\ -172.35 \\ 1.4369 \\ -166.8 \end{bmatrix},$$

$$\hat{C}_2 = \begin{bmatrix} 0.2093 & 0.5365 & 0.8714 & -0.0000 \\ 0.0965 & -0.0207 & 0.3118 & 0.0000 \end{bmatrix}.$$

In our simulation, we use a uniform distributed random disturbance input signal $w(t)$ with maximum value 2. The mode transition of the fault estimator during the simulation is depicted in Figure 7.2 with initial mode 2. Mode 1 represents controller $\mathscr{G}_1$ while mode 2 for $\mathscr{G}_2$. Histories of the residual signals $r_{s_1}(t)$ and $r_{s_2}(t)$ are shown in Figure 7.3 and Figure 7.4, respectively. These results demonstrate that the designed fault estimator meets the performance requirement.

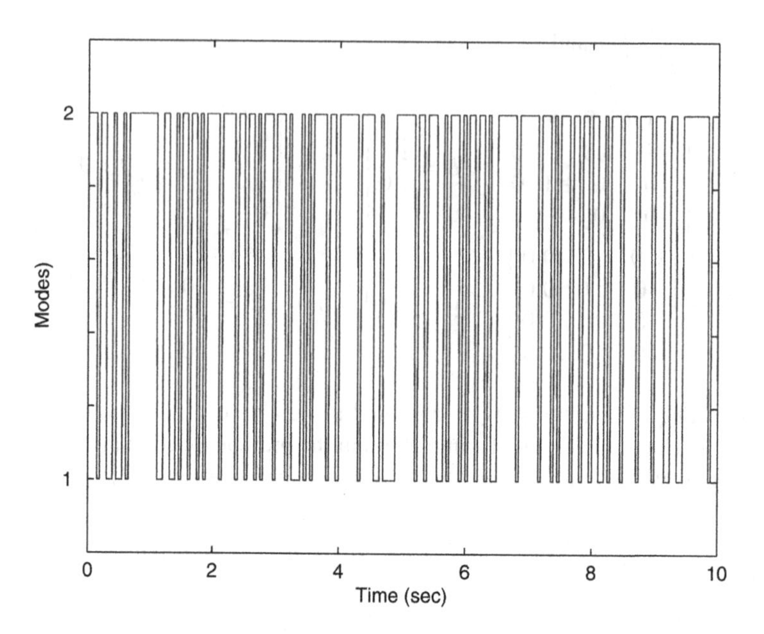

Fig. 7.2 Mode transitions

7.4 Conclusion

In this chapter, a technique of designing a delay-dependant dynamic fault estimator for an uncertain NCS with random communication network-induced delays and data packet dropouts has been proposed. The network-induced delays and data packet dropouts are regarded as input delays. The Lyapunov–Razumikhin method has been employed to derive a fault estimator for this class of systems. Sufficient conditions for the existence of such a fault estimator for this class of NCSs are derived. We finally use a numerical example to demonstrate the effectiveness of this methodology.

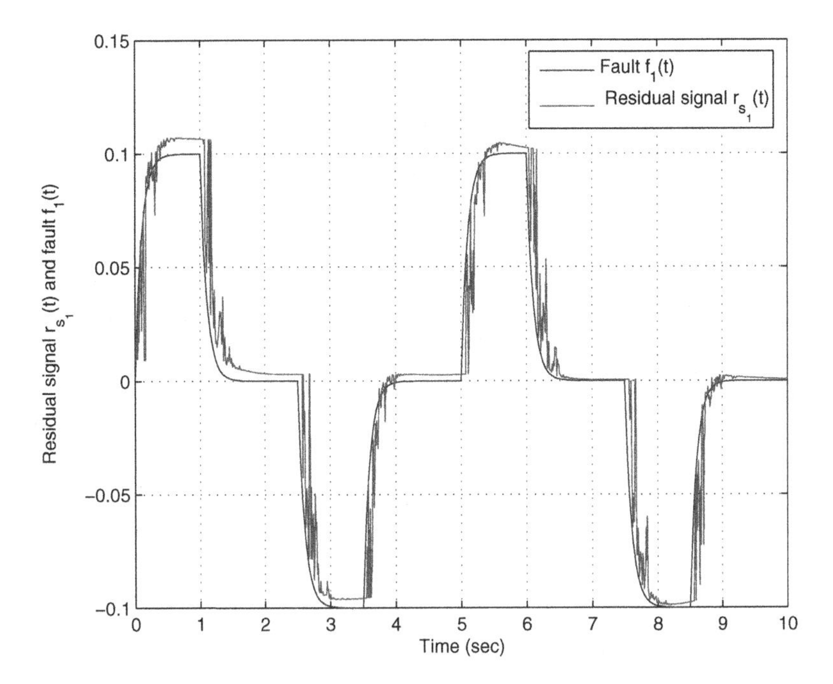

Fig. 7.3 Residual signals $r_{s_1}(t)$ and $f_1(t)$

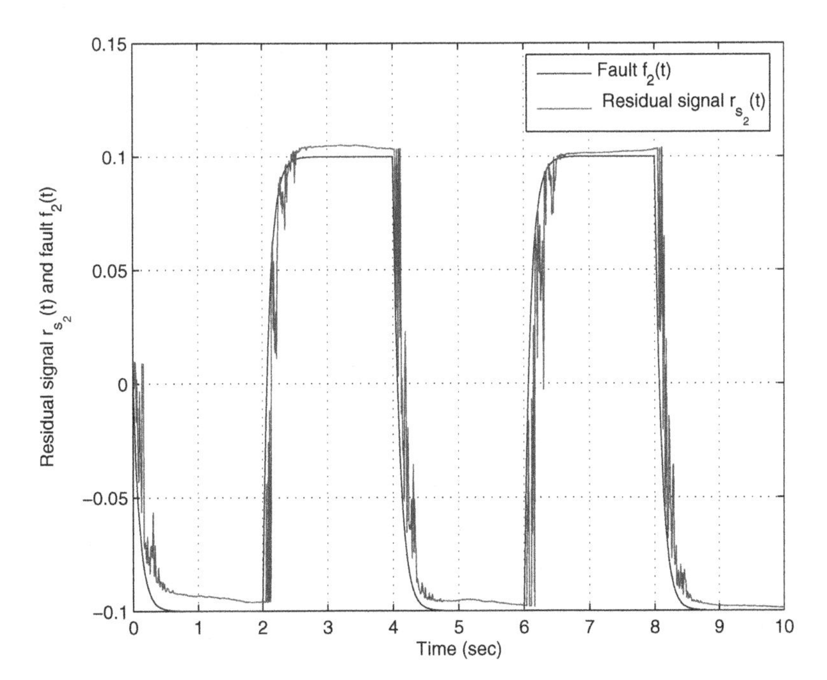

Fig. 7.4 Residual signals $r_{s_2}(t)$ and $f_2(t)$

Part II
Nonlinear Uncertain Networked Control Systems

Chapter 8
Takagi–Sugeno Fuzzy Control System

A nonlinear dynamic system can usually be represented by a set of nonlinear differential equations of the form

$$\dot{x} = f(x,u), \tag{8.1}$$

where $f(\bullet)$ is a nonlinear vector function, x is the state vector, and u is the control input.

In the rest parts of the book, we approximate the nonlinear plant (8.1) by a T-S fuzzy model [62]. This fuzzy modeling is simple and natural. The system dynamics are captured by a set of fuzzy implications which characterize local relations in the state space. The main feature of a T-S fuzzy model is to express the local dynamics of each fuzzy implication (rule) by a linear system model. The overall fuzzy model of the system is achieved by fuzzy "blending" of the linear system models.

Specifically, the T-S fuzzy system is described by the fuzzy IF-THEN rules , which locally represent linear input-output relations of a system. The ith rules of the T-S fuzzy models are of the following forms:

Plant Rule i:

IF $v_1(t)$ is M_{i1} and $\cdots$ and $v_p(t)$ is M_{ip}, THEN

$$\begin{aligned} \dot{x}(t) &= A_i x(t) + B_i u(t), x(0) = 0 \\ y(t) &= C_i x(t) + D_i u(t) \end{aligned} \tag{8.2}$$

where $i = 1, \cdots, r$, r is the number of fuzzy rules; $v_k(t)$ are premise variables, M_{ik} are fuzzy sets, $k = 1, \cdots, p$, p is the number of premise variables; $x(t) \in \Re^n$ is the state, $u(t) \in \Re^m$ is the control input, $y(t) \in \Re^l$ is the output, the matrices A_i, B_i, C_i, D_i are of appropriate dimension.

For any given $x(t)$ and $u(t)$, by using a centre-average defuzzifier, product inference and singleton fuzzifier, the local models can be integrated into a global nonlinear model:

$$\begin{aligned} \dot{x}(t) &= \sum_{i=1}^{r} \mu_i(v(t))[A_i x(t) + B_i u(t)] \\ y(t) &= \sum_{i=1}^{r} \mu_i(v(t))[C_i x(t) + D_i u(t)] \end{aligned} \tag{8.3}$$

where

D. Huang and S.K. Nguang: Robust Ctrl. for Uncertain Networked Ctrl. Sys., LNCIS 386, pp. 87–91.
springerlink.com

$$v(t) = [v_1(t), v_2(t), \cdots, v_p(t)]^T,$$

$$\omega_i(v(t)) = \prod_{k=1}^{p} M_{ik}(v_k(t)),\ \omega_i(v(t)) \geq 0,\ \sum_{i=1}^{r} \omega_i(v(t)) > 0,$$

$$\mu_i(v(t)) = \frac{\omega_i(v(t))}{\sum_{i=1}^{r} \omega_i(v(t))},\ \mu_i(v(t)) \geq 0,\ \sum_{i=1}^{r} \mu_i(v(t)) = 1.$$

Here, $M_{ik}(v_k(t))$ denote the grade of membership of $v_k(t)$ in M_{ik}. Figure 8.1 shows the structural diagram of the T-S fuzzy system.

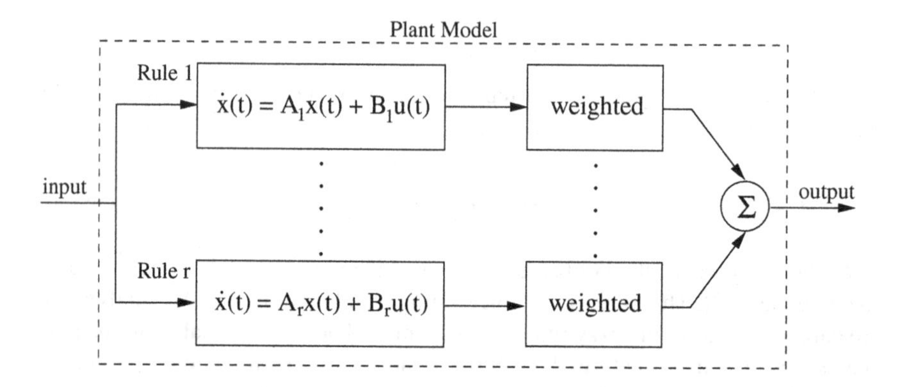

Fig. 8.1 The T-S type fuzzy system

8.1 Takagi-Sugeno Fuzzy Modeling

Two methods have been commonly applied in T-S fuzzy modeling. One is the T-S fuzzy model identification using input-output data, while the other is the T-S fuzzy model construction using the idea of sector nonlinearity. In this book, we only consider the latter method.

Consider the following nonlinear system:

$$\dot{x}_i(t) = \sum_{j=1}^{n} f_{ij}(x(t))x_j(t) + \sum_{k=1}^{m} g_{ik}(x(t))u_k(t), \tag{8.4}$$

where n and m are, respectively, the numbers of state variables and inputs. $x(t) = [x_1(t)\ \cdots\ x_n(t)]$ is the state vector and $u(t) = [u_1(t)\ \cdots\ u_n(t)]$ is the input vector. $f_{ij}(x(t))$ and $g_{ik}(x(t))$ are functions of $x(t)$. To obtain a T-S fuzzy model, we find the minimum and maximum values of $f_{ij}(x(t))$ and $g_{ik}(x(t))$,

$$a_{ij1} = \max_{x(t)} \{f_{ij}(x(t))\}, \; a_{ij2} = \min_{x(t)} \{f_{ij}(x(t))\},$$
$$b_{ik1} = \max_{x(t)} \{g_{ik}(x(t))\}, \; b_{ik2} = \max_{x(t)} \{g_{ik}(x(t))\},$$

where $x(t) \in [m_l, m_u]$. m_l and m_u are lower limit and upper limit of $x(t)$, respectively.

By using these variables, f_{ij} and g_{ik} can be represented as:

$$f_{ij}(x(t)) = \sum_{l=1}^{2} h_{ijl}(x(t))a_{ijl},$$
$$g_{ik}(x(t)) = \sum_{l=1}^{2} v_{ikl}(x(t))b_{ikl},$$

where $\sum_{l=1}^{2} h_{ijl}(x(t)) = 1$ and $\sum_{l=1}^{2} v_{ikl}(x(t)) = 1$. The membership functions are assigned as follows:

$$h_{ij1}(x(t)) = \frac{f_{ij}(x(t)) - a_{ij2}}{a_{ij1} - a_{ij2}}, \; h_{ij2}(x(t)) = \frac{a_{ij1} - f_{ij}(x(t))}{a_{ij1} - a_{ij2}},$$
$$g_{ik1}(x(t)) = \frac{g_{ik}(x(t)) - b_{ik2}}{b_{ik1} - b_{ik2}}, \; g_{ik2}(x(t)) = \frac{b_{ik1} - g_{ij1}}{b_{ik1} - b_{ik2}}.$$

By using the fuzzy mode representation, (8.4) can be rewritten as

$$\dot{x}(t) = \sum_{j=1}^{n} \sum_{l=1}^{2} h_{ijl}(x(t))a_{ijl}x(t) + \sum_{k=1}^{m} \sum_{l=1}^{2} g_{ik}(x(t))v_{ikl}(x(t))b_{ikl}u(t). \tag{8.5}$$

The following example is used to illustrate the T-S fuzzy modeling procedure.

Example 8.1. Nonlinear Mass-Spring-Damper System

Consider a nonlinear mass-spring-damper mechanical system with a nonlinear spring:

$$\begin{aligned} \dot{x}_1(t) &= -0.1125x_1(t) - 0.02x_2(t) - 0.67x_2^3(t) + u(t) \\ \dot{x}_2(t) &= x_1(t) \end{aligned} \tag{8.6}$$

where $x_2(t)$ is the spring's displacement and $x_1(t) = \dot{x}_2(t)$. The term $-0.67x_2^3$ is due to the nonlinearity of the spring. The spring is attached to a fixed wall, therefore the spring's displacement $x_2(t)$ is physically constrained by the length of the spring and the wall. The length of the spring could be any value, and we assume $x_2(t) \in [-1 \; 1.5]$. The lower limit is the minimum length that the spring can be compressed. The concept of sector nonlinearity [124] is employed to construct an exact T-S fuzzy model for the mass-spring-damper system. Using the fact that $x_2(t) \in [-1 \; 1.5]$, this nonlinear term can be expressed as

$$-0.67x_2^3(t) = -h_1(x_2(t))[0 \cdot x_2(t)] - h_2(x_2(t))[1.5075 \cdot x_2(t)],$$

where $h_1(x_2(t)) = 1 - \frac{x_2^2(t)}{2.25}$ and $h_2(x_2(t)) = \frac{x_2^2(t)}{2.25}$.

Using $h_1(x_2(t))$ and $h_2(x_2(t))$, we obtain the following T-S fuzzy model which exactly represents (8.6) under the assumption on the limits of the state variable $x_2(t) \in [-1\ 1.5]$:

$$\dot{x}(t) = \Sigma_{i=1}^{2} h_i(x_2(t))A_i x(t) + \Sigma_{i=1}^{2} h_i(x_2(t))B_i u(t),$$

where

$$x(t) = \begin{bmatrix} x_1(t) \\ x_2(t) \end{bmatrix},\ A_1 = \begin{bmatrix} -0.1125 & -0.02 \\ 1 & 0 \end{bmatrix},$$
$$A_2 = \begin{bmatrix} -0.1125 & -1.5275 \\ 1 & 0 \end{bmatrix},\ B_1 = B_2 = \begin{bmatrix} 1 \\ 0 \end{bmatrix}.$$

8.2 Takagi-Sugeno Fuzzy Controller

For the nonlinear plant represented by (8.3), the fuzzy controller is designed to share the same IF parts with the plants as follows:

Controller Rule i:

IF $v_1(t)$ is M_{i1} and $\cdots$ and $v_p(t)$ is M_{ip}, THEN

$$u(t) = F_i x(t). \tag{8.7}$$

Hence, the overall controller is represented by:

$$u(t) = \sum_{i=1}^{r} \mu_i(v(t))F_i x(t). \tag{8.8}$$

The block diagram of the TS fuzzy controller is given in Figure 8.2.

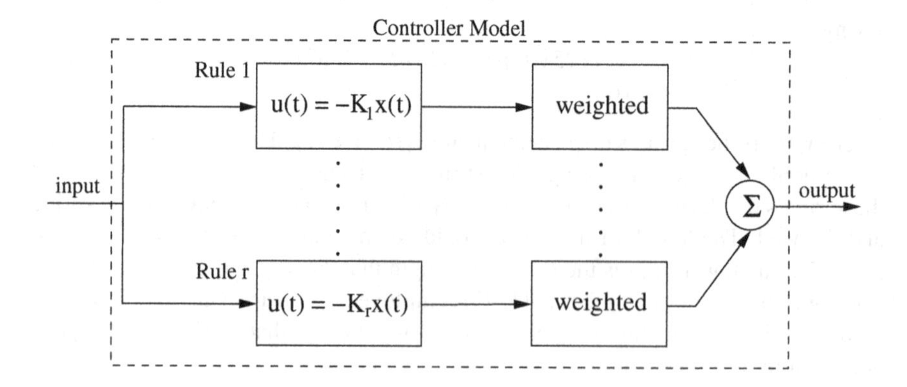

Fig. 8.2 The T-S type fuzzy controller

Note that the resulting controller (8.8) is nonlinear in general since the coefficients of the controller depend nonlinearly on the system input and output fuzzy weights. Moreover, the resulting fuzzy controller (8.8) could be represented as a particular form of a gain scheduled controller where the gains are varied as a function of operating conditions. The T-S type fuzzy control scheme has a major advantage over the existing crisp gain scheduling scheme. That is, it provides a general method for the interpolation of available local control law into an overall gain scheduling control law.

Chapter 9
State Feedback Control for Uncertain Nonlinear Networked Control Systems

9.1 Introduction

In this chapter, our concern is to consider a class of nonlinear uncertain NCSs with sensors and actuators connected to a controller via two communication networks in the continuous-time domain. Linear feedback control techniques can be utilized as in the case of feedback stabilization. The design procedure is stated as follows. First the nonlinear plant is represented by a T-S fuzzy model. In this type of fuzzy model, local dynamics in different state space regions are represented by linear models. The overall model of the system is achieved by "blending" of these linear models. Then for each local linear model, a linear feedback controller is to be designed. The resulting overall controller, which is nonlinear in general, is again a fuzzy blending of each individual linear controller.

Two Markov processes are used to model the network-induced delays which randomly occurs in both of these two networks. A set of linear local controllers for each plant of the T-S model are then designed based on the Lyapunov–Razumikhin method. Sufficient conditions for the existence of a mode-dependent for this class of nonlinear NCSs are derived.

In the system setup, modeling procedure introduced in Chapter 2 is applied in this chapter.

9.2 Problem Formulation and Preliminaries

A fuzzy dynamic model has been proposed by Takagi and Sugeno [62] to represent local linear input/output relations of nonlinear systems. A class of uncertain nonlinear systems under consideration in this chapter is described by the following IF-THEN rules and the ith rule has been shown as follows:

Plant Rule i:
IF $v_1(t)$ is M_{i1} and $\cdots$ and $v_p(t)$ is M_{ip},
THEN

D. Huang and S.K. Nguang: Robust Ctrl. for Uncertain Networked Ctrl. Sys., LNCIS 386, pp. 93–105.
springerlink.com

$$\dot{x}(t) = (A_i + \Delta A_i)x(t) + (B_i + \Delta B_i)u(t), \tag{9.1}$$

where $i \in \mathscr{I}_R = \{1, \cdots, r\}$, r is the number of fuzzy rules; $v_k(t)$ are premise variables, M_{ik} are fuzzy sets, $k = 1, \cdots, p$, p is the number of premise variables; $x(t) \in \Re^n$ and $u(t) \in \Re^m$ denote state and control input, respectively. Matrices $A_i \in \Re^{n\times n}$ and $B_i \in \Re^{n\times m}$ are known system matrices. Matrices ΔA_i and ΔB_i represent the uncertainties in the system and satisfy the following assumption.

Assumption 9.1. The parameter uncertainties considered here are norm-bounded, in the form

$$\left[\Delta A_i \ \Delta B_i\right] = H_i F_i(t) \left[E_{1i} \ E_{2i}\right], \tag{9.2}$$

where H_i, E_{1i} and E_{2i} are known real constant matrices of appropriate dimensions, and $F_i(t)$ is an unknown matrix function with Lebesgue-measurable elements and satisfies $F_i(t)^T F_i(t) \leq I$, in which I is the identity matrix of appropriate dimension.

By using a center-average defuzzifier, product inference and singleton fuzzifier, the local models can be integrated into a global nonlinear model:

$$\dot{x}(t) = \sum_{i=1}^{r} \mu_i(v(t))[(A_i + \Delta A_i)x(t) + (B_i + \Delta B_i)u(t)], \tag{9.3}$$

where

$$v(t) = [v_1(t), v_2(t), \cdots, v_p(t)]^T,$$

$$\omega_i(v(t)) = \prod_{k=1}^{p} M_{ik}(v_k(t)),\ \omega_i(v(t)) \geq 0,\ \sum_{i=1}^{r} \omega_i(v(t)) > 0,$$

$$\mu_i(v(t)) = \frac{\omega_i(v(t))}{\sum_{i=1}^{r} \omega_i(v(t))},\ \mu_i(v(t)) \geq 0,\ \sum_{i=1}^{r} \mu_i(v(t)) = 1.$$

Here, $M_{ik}(v_k(t))$ denote the grade of membership of $v_k(t)$ in M_{ik} and $v(t)$ is the premise vector.

Hereinafter, we apply the same procedure as presented in Chapter 2 to model the network-induced delay and data packet dropouts by using two Markov processes.

In this chapter, we consider a nonlinear NCS of which the plant is described by the T-S model (9.3). The setup of the overall control system is depicted in Figure 9.1, where $\tau(\eta_1(t), t) \geq 0$ is the random time-delay from sensor to controller and $\rho(\eta_2(t), t) \geq 0$ is the random time-delay from controller to actuator. These delays are assumed to be upper bounded.

It should be noted that in this setup, premise vector $v(t)$ is connected to the actuator via point-to-point architecture, which is immune to network-induced delays.

Therefore, following the modeling procedure presented in Chapter 2, for the nonlinear plant represented by (9.3), the fuzzy state feedback controller at time t is inferred as follows:

$$u(t) = \sum_{i=1}^{r} \mu_i(v(t)) K_i(\eta_1(t), \eta_2(t)) x(t - \tau(\eta_1(t), t)), \tag{9.4}$$

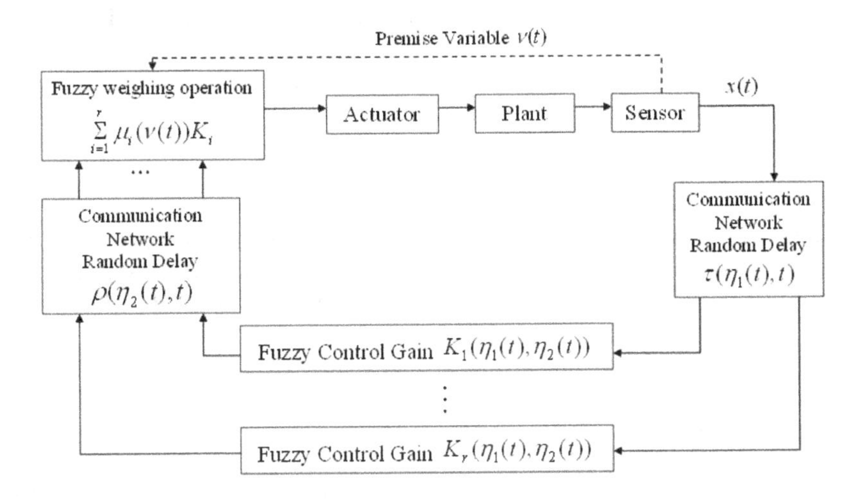

Fig. 9.1 A fuzzy NCS with random delays

where $K_i(\eta_1(t), \eta_2(t))$ in each plant rule is a local mode-dependant controller gain to be designed. The plant is rewritten as:

$$\dot{x}(t) = \sum_{i=1}^{r} \mu_i(v(t))[(A_i + \Delta A_i)x(t) + (B_i + \Delta B_i)u(t - \rho(\eta_2(t),t))]. \tag{9.5}$$

Substituting (9.4) into (9.5) yields

$$\begin{aligned}\dot{x}(t) = \sum_{i=1}^{r}\sum_{j=1}^{r} \mu_i(v(t))\mu_j(v(t))\Big[[A_i + \Delta A_i]x(t) \\ +(B_i + \Delta B_i)K_j(\eta_1(t),\eta_2(t))x(t - \tau(\eta_1(t),t) - \rho(\eta_2(t),t))\Big].\end{aligned} \tag{9.6}$$

The aim of this chapter is to design a fuzzy state feedback controller of the form (9.4) such that the following inequality holds:

$$\begin{aligned}&\mathbf{E}\Big[\max_{\eta_1(t)\in\mathscr{S},\eta_2(t)\in\mathscr{W}} \tilde{A}V(x(t),\eta_1(t),\eta_2(t),t)\Big] \\ &\leq -\zeta\mathbf{E}\Big[\max_{\eta_1(t)\in\mathscr{S},\eta_2(t)\in\mathscr{W}} V(x(t),\eta_1(t),\eta_2(t),t)\Big],\end{aligned} \tag{9.7}$$

provided $x = \{x(\xi) : t - 2\chi \leq \xi \leq t\}$ satisfying:

$$\begin{aligned}&\mathbf{E}\Big[\min_{\eta_1(t)\in\mathscr{S},\eta_2(t)\in\mathscr{W}} V(x(\xi),\eta_1(\xi),\eta_2(\xi),\xi)\Big] \\ &< \delta\mathbf{E}\Big[\max_{\eta_1(t)\in\mathscr{S},\eta_2(t)\in\mathscr{W}} V(x(t),\eta_1(t),\eta_2(t),t)\Big],\end{aligned} \tag{9.8}$$

for all $t-2\chi \leq \xi \leq t$. Here $\tilde{A}$ denotes the weak infinitesimal operator and $\mathbf{E}[\cdot]$ stands for the mathematical expectation.

Then the system (9.6) is said to achieve stochastic stability with Markovian jumps.

In this chapter, we assume $u(t)=0$ before the first control signal reaches the plant. In this system setup, sensor is clock-driven while controller and actuator are event-driven.

From here, $\mu_i(v(t))$ and $\mu_j(v(t))$ are denoted as μ_i and μ_j respectively for the convenience of notations. In the symmetric block matrices, we use (*) as an ellipsis for terms that are induced by symmetry. $K_i(\eta_1(t),\eta_2(t))$ is denoted as $K_i(\iota,\kappa)$ if $\eta_1(t)=\iota$ and $\eta_2(t)=\kappa$.

9.3 Main Result

The following theorem provides sufficient conditions for the existence of a mode-dependent state feedback controller for the system (9.6).

Theorem 9.1. *Consider the system (9.6) satisfying Assumption 9.1. Given a positive scalar $\hbar(\iota,\kappa)$, if there exist a symmetric matrix $P(\iota,\kappa)>0$, a matrix $K(\iota,\kappa)$ and positive scalars $\beta_{1_{\iota\kappa}}$, $\beta_{2_{\iota\kappa}}$, $\beta_{3_{\iota\kappa}}$, $\varepsilon_{1_{ii\iota\kappa}}$, $\varepsilon_{2_{ij_{\iota\kappa}}}$, $\varepsilon_{3_{ij_{\iota\kappa}}}$, and $\varepsilon_{ij_{\iota\kappa}}$ such that the following inequalities hold for all $\iota\in\mathscr{S}$ and $\kappa\in\mathscr{W}$:*

$$\begin{bmatrix} \Omega_i(\iota,\kappa) & (*)^T & (*)^T \\ H_i^T P(\iota,\kappa) & -\varepsilon_{ii_{\iota\kappa}} I & (*)^T \\ \varepsilon_{ii_{\iota\kappa}}(E_{1i}+E_{2i}K_i(\iota,\kappa)) & 0 & -\varepsilon_{ii_{\iota\kappa}} I \end{bmatrix} < 0, \quad \text{for } i\in\mathscr{I}_R \tag{9.9}$$

$$\begin{bmatrix} \Omega_{ij}(\iota,\kappa) & (*)^T & (*)^T \\ H_i^T P(\iota,\kappa) & -2\varepsilon_{ij_{\iota\kappa}} I & (*)^T \\ \varepsilon_{ij_{\iota\kappa}}\Upsilon_{ij} & 0 & -2\varepsilon_{ij_{\iota\kappa}} I \end{bmatrix} < 0, \quad \text{for } i<j<r \tag{9.10}$$

$$\begin{bmatrix} -\beta_{1_{\iota\kappa}} P(\iota,\kappa) & (*)^T & (*)^T \\ A_i^T P(\iota,\kappa) & -I+\varepsilon_{1_{ii\iota\kappa}} E_{1i}^T E_{1i} & (*)^T \\ H_i^T P(\iota,\kappa) & 0 & -\varepsilon_{1_{ii\iota\kappa}} I \end{bmatrix} < 0, \quad \text{for } i\in\mathscr{I}_R \tag{9.11}$$

$$\begin{bmatrix} -\beta_{2_{\iota\kappa}} P(\iota,\kappa) & (*)^T & (*)^T & (*)^T \\ K_j^T(\iota,\kappa) B_i P(\iota,\kappa) & -I & (*)^T & (*)^T \\ H_i^T P(\iota,\kappa) & 0 & -\varepsilon_{2_{ij_{\iota\kappa}}} I & (*)^T \\ 0 & \varepsilon_{2_{ij_{\iota\kappa}}} E_{2i} K_j(\iota,\kappa) & 0 & -\varepsilon_{2_{ij_{\iota\kappa}}} I \end{bmatrix} < 0, \quad \text{for } \{i,j\}\in\mathscr{I}_R\times\mathscr{I}_R \tag{9.12}$$

where

$$\begin{aligned}
\Omega_i(\iota,\kappa) &= A_i^T P(\iota,\kappa) + P(\iota,\kappa)A_i + K_i^T B_i^T P(\iota,\kappa) + P(\iota,\kappa)B_iK_i \\
&\quad + \sum_{\wp=1}^{s} \lambda_{\iota\wp}P(\wp,\kappa) + \sum_{\ell=1}^{w} \pi_{\kappa\ell}P(\iota,\ell) + \hbar(\iota,\kappa)(\beta_{1_{\iota\kappa}} + 3\beta_{2_{\iota\kappa}})P(\iota,\kappa), \\
\Omega_{ij}(\iota,\kappa) &= \frac{1}{2}[A_i^T P(\iota,\kappa) + P(\iota,\kappa)A_i + A_j^T P(\iota,\kappa) + P(\iota,\kappa)A_j \\
&\quad + K_j^T B_i^T P(\iota,\kappa) + P(\iota,\kappa)B_iK_j + K_i^T B_j^T P(\iota,\kappa) + P(\iota,\kappa)B_jK_i] \\
&\quad + \sum_{\wp=1}^{s} \lambda_{\iota\wp}P(\wp,\kappa) + \sum_{\ell=1}^{w} \pi_{\kappa\ell}P(\iota,\ell) + \hbar(\iota,\kappa)(\beta_{1_{\iota\kappa}} + 3\beta_{2_{\iota\kappa}})P(\iota,\kappa), \\
\Upsilon_{ij}(\iota,\kappa) &= E_{1i} + E_{2i}K_j(\iota,\kappa) + E_{1j} + E_{2j}K_i(\iota,\kappa),
\end{aligned}$$

then the system (9.6) is said to achieve stochastic stability via the controller (9.4) for all delays $\tau(\iota,t)$ and $\rho(\kappa,t)$ satisfying

$$0 \le \tau(\iota,t) + \rho(\kappa,t) \le \hbar(\iota,\kappa).$$

Proof. Note that for each $\eta_1(t) = \iota \in \mathcal{S}$ and $\eta_2(t) = \kappa \in \mathcal{W}$,

$$\begin{aligned}
&x(t - (\tau(\iota,t) + \rho(\kappa,t))) \\
&= x(t) - \int_{-(\tau(\iota,t)+\rho(\kappa,t))}^{0} \dot{x}(t+\theta)d\theta \\
&= \sum_{i=1}^{r}\sum_{j=1}^{r} \mu_i\mu_j \Big\{ x(t) - \int_{-(\tau(\iota,t)+\rho(\kappa,t))}^{0} \Big[[A_i + \Delta A_i]x(t+\theta) \\
&\quad + (B_i + \Delta B_i)K_j(\iota,\kappa)x(t - \tau(\iota,t) - \rho(\kappa,t) + \theta) \Big] d\theta \Big\}.
\end{aligned} \tag{9.13}$$

Using (9.13), the closed-loop system (9.6) can be rewritten as:

$$\begin{aligned}
&\dot{x}(t) \\
&= \sum_{i=1}^{r}\sum_{j=1}^{r}\sum_{k=1}^{r}\sum_{l=1}^{r} \mu_i\mu_j\mu_k\mu_l \Big\{ [A_i + \Delta A_i + (B_i + \Delta B_i)K_j(\iota,\kappa)]x(t) \\
&\quad - (B_i + \Delta B_i)K_j(\iota,\kappa) \int_{-(\tau(\iota,t)+\rho(\kappa,t))}^{0} \Big[[A_k + \Delta A_k]x(t+\theta) \\
&\quad + (B_k + \Delta B_k)K_l(\iota,\kappa)x(t - \tau(\iota,t) - \rho(\kappa,t) + \theta) \Big] d\theta \Big\}.
\end{aligned} \tag{9.14}$$

For the sake of notation simplification, $K_i(\iota,\kappa)$ will be denoted as K_i in the rest of this chapter. We also define $\tau(\iota,t) + \rho(\kappa,t) = \chi(t)$.

Select a Lyapunov function candidate as

$$V(x(t),\eta_1(t),\eta_2(t),t) = x^T(t)P(\eta_1(t),\eta_2(t))x(t), \tag{9.15}$$

where $P(\eta_1(t),\eta_2(t))$ is the positive symmetric matrix. It follows

$$\alpha_1\|x(t)\|^2 \le V(x(t),\eta_1(t),\eta_2(t),t) \le \alpha_2\|x(t)\|^2, \tag{9.16}$$

where $\alpha_1 = \lambda_{min}(P(\eta_1(t),\eta_2(t)))$ and $\alpha_2 = \lambda_{max}(P(\eta_1(t),\eta_2(t)))$.

The weak infinitesimal operator $\tilde{A}$ can be considered as the derivative of the function of $V(x(t), \eta_1(t), \eta_2(t), t)$ along the trajectory of the joint Markov process $\{x(t), \eta_1(t), \eta_2(t), t \geq 0\}$ at the point $\{x(t), \eta_1(t) = \iota, \eta_2(t) = \kappa\}$ at time t;

$$\begin{aligned} &\tilde{A}V(x(t), \eta_1(t), \eta_2(t), t) \\ &= \frac{\partial V(\cdot)}{\partial t} + \dot{x}^T(t)\frac{\partial V(\cdot)}{\partial x}\bigg|_{\eta_1=\iota, \eta_2=\kappa} + \sum_{\wp=1}^{s} \lambda_{\iota\wp} V(x(t), \wp, \kappa, t) + \sum_{\ell=1}^{w} \pi_{\kappa\ell} V(x(t), \iota, \ell, t). \end{aligned} \tag{9.17}$$

It follows from (9.17) that

$$\begin{aligned} &\tilde{A}V(x(t), \iota, \kappa, t) \\ &= \dot{x}^T(t)P(\iota,\kappa)x(t) + x^T(t)P(\iota,\kappa)\dot{x}(t) + \sum_{\wp=1}^{s} \lambda_{\iota\wp} x^T(t)P(\wp,\kappa)x(t) + \sum_{\ell=1}^{w} \pi_{\kappa\ell} x^T(t)P(\iota,\ell)x(t) \\ &= \sum_{i=1}^{r}\sum_{j=1}^{r}\sum_{k=1}^{r}\sum_{l=1}^{r} \mu_i\mu_j\mu_k\mu_l \Big\{ x^T(t)\Big[(A_i + \Delta A_i)^T P(\iota,\kappa) + P(\iota,\kappa)(A_i + \Delta A_i) \\ &\quad + K_j^T(B_i + \Delta B_i)^T P(\iota,\kappa) + P(\iota,\kappa)(B_i + \Delta B_i)K_j + \sum_{\wp=1}^{s} \lambda_{\iota\wp} P(\wp,\kappa) + \sum_{\ell=1}^{w} \pi_{\kappa\ell} P(\iota,\ell)\Big] x(t) \\ &\quad - 2\int_{-\chi(t)}^{0} \Big[x^T(t)P(\iota,\kappa)(B_i + \Delta B_i)K_j\Big[[A_k + \Delta A_k]x(t+\theta) \\ &\qquad + (B_k + \Delta B_k)K_l x(t - \chi(t) + \theta)\Big]\Big] d\theta \Big\} \\ &\leq x^T(t)\mathscr{M}_{\iota\kappa}(\chi(t), \delta)x(t) + \int_{-\chi(t)}^{0} \Big[\beta_{1_{\iota\kappa}} x^T(t+\theta)P(\iota,\kappa)x(t+\theta) \\ &\quad + \beta_{2_{\iota\kappa}} x^T(t - \chi(t) + \theta)P(\iota,\kappa)x(t - \chi(t) + \theta) - (\beta_{1_{\iota\kappa}} + \beta_{2_{\iota\kappa}})\delta x^T(t)P(\iota,\kappa)x(t)\Big] d\theta, \end{aligned}$$

where $\mathscr{M}_{\iota\kappa}(\cdot,\cdot)$ is given by:

$$\begin{aligned} &\mathscr{M}_{\iota\kappa}(\chi(t), \delta) \\ &= \sum_{i=1}^{r}\sum_{j=1}^{r}\sum_{k=1}^{r}\sum_{l=1}^{r} \mu_i\mu_j\mu_k\mu_l \Big\{ (A_i + \Delta A_i)^T P(\iota,\kappa) + P(\iota,\kappa)(A_i + \Delta A_i) \\ &\quad + K_j^T(B_i + \Delta B_i)^T P(\iota,\kappa) + P(\iota,\kappa)(B_i + \Delta B_i)K_j + \sum_{\wp=1}^{s} \lambda_{\iota\wp} P(\wp,\kappa) + \sum_{\ell=1}^{w} \pi_{\kappa\ell} P(\iota,\ell) \\ &\quad + \chi(t)\Big[\beta_{1_{\iota\kappa}}^{-1} P(\iota,\kappa)(B_i + \Delta B_i)K_j(A_k + \Delta A_k)P^{-1}(\iota,\kappa)(A_k + \Delta A_k)^T K_j^T(B_i + \Delta B_i)^T P(\iota,\kappa) \\ &\quad + \beta_{2_{\iota\kappa}}^{-1} P(\iota,\kappa)(B_i + \Delta B_i)K_j(B_k + \Delta B_k)K_l P^{-1}(\iota,\kappa)K_l^T(B_k + \Delta B_k)^T K_j^T(B_i + \Delta B_i)^T P(\iota,\kappa) \\ &\quad + (\beta_{1_{\iota\kappa}} + \beta_{2_{\iota\kappa}})\delta P(\iota,\kappa)\Big]\Big\}. \end{aligned}$$

In this chapter, the time-delays are assumed to be bounded, hence, $\chi(t)$ can also be assumed to be bounded, that is, $\chi(t) \leq \hbar(\iota,\kappa)$, where $\hbar(\iota,\kappa)$ are constants given in the theorem. Using this fact, we learn that

$$\mathscr{M}_{\iota\kappa}(\chi(t), \delta) \leq \mathscr{M}_{\iota\kappa}(\hbar(\iota,\kappa), \delta).$$

If (9.11)-(9.12) hold, by applying Lemma A.2 and Schur complement, we get:

$$(A_k + \Delta A_k)P^{-1}(\iota,\kappa)(A_k + \Delta A_k)^T < \beta_{1_{\iota\kappa}} P^{-1}(\iota,\kappa), \qquad (9.18)$$
$$(B_k + \Delta B_k)K_l P^{-1}(\iota,\kappa)K_l^T (B_k + \Delta B_k)^T < \beta_{2_{\iota\kappa}} P^{-1}(\iota,\kappa). \qquad (9.19)$$

Using (9.18)-(9.19), $\mathscr{M}_{\iota\kappa}(\hbar(\iota,\kappa),\delta)$ becomes:

$$\sum_{i=1}^{r}\sum_{j=1}^{r}\mu_i\mu_j\Big[(A_i+\Delta A_i)^T P(\iota,\kappa) + P(\iota,\kappa)(A_i+\Delta A_i)$$
$$+K_j^T(B_i+\Delta B_i)^T P(\iota,\kappa) + P(\iota,\kappa)(B_i+\Delta B_i)K_j + \sum_{\wp=1}^{s}\lambda_{\iota\wp}P(\wp,\kappa) + \sum_{\ell=1}^{w}\pi_{\kappa\ell}P(\iota,\ell)$$
$$+2\hbar(\iota,\kappa)\beta_{2_{\iota\kappa}}P(\iota,\kappa) + \hbar(\iota,\kappa)(\beta_{1_{\iota\kappa}} + \beta_{2_{\iota\kappa}})\delta P(\iota,\kappa)\Big]$$
$$= \sum_{i=1}^{r}\mu_i^2\Big[(A_i+\Delta A_i)^T P(\iota,\kappa) + P(\iota,\kappa)(A_i+\Delta A_i)$$
$$+K_i^T(B_i+\Delta B_i)^T P(\iota,\kappa) + P(\iota,\kappa)(B_i+\Delta B_i)K_i + \sum_{\wp=1}^{s}\lambda_{\iota\wp}P(\wp,\kappa) + \sum_{\ell=1}^{w}\pi_{\kappa\ell}P(\iota,\ell)$$
$$+2\hbar(\iota,\kappa)\beta_{2_{\iota\kappa}}P(\iota,\kappa) + \hbar(\iota,\kappa)(\beta_{1_{\iota\kappa}} + \beta_{2_{\iota\kappa}})\delta P(\iota,\kappa)\Big]$$
$$+2\sum_{i=1}^{r}\sum_{i<j}^{r}\mu_i\mu_j\Big[\frac{1}{2}[(A_i+\Delta A_i)^T P(\iota,\kappa) + P(\iota,\kappa)(A_i+\Delta A_i)$$
$$+(A_j+\Delta A_j)^T P(\iota,\kappa) + P(\iota,\kappa)(A_j+\Delta A_j) + K_j^T(B_i+\Delta B_i)^T P(\iota,\kappa) + P(\iota,\kappa)(B_i+\Delta B_i)K_j$$
$$+K_i^T(B_j+\Delta B_j)^T P(\iota,\kappa) + P(\iota,\kappa)(B_j+\Delta B_j)K_i] + \sum_{\wp=1}^{s}\lambda_{\iota\wp}P(\wp,\kappa) + \sum_{\ell=1}^{w}\pi_{\kappa\ell}P(\iota,\ell)$$
$$+2\hbar(\iota,\kappa)\beta_{2_{\iota\kappa}}P(\iota,\kappa) + \hbar(\iota,\kappa)(\beta_{1_{\iota\kappa}} + \beta_{2_{\iota\kappa}})\delta P(\iota,\kappa)\Big].$$

Hence, if (9.9) and (9.10) hold, it is not hard to see that $\mathscr{M}_{\iota\kappa}(\hbar(\iota,\kappa),1) < 0$ for $\delta = 1$.

Thus,

$$\tilde{A}V(x(t),\eta_1(t),\eta_2(t),t)$$
$$\leq -\alpha x^T(t)x(t) + \int_{-\chi(t)}^{0}\Big[\beta_{1_{\iota\kappa}}x^T(t+\theta)P(\iota,\kappa)x(t+\theta)$$
$$+\beta_{2_{\iota\kappa}}x^T(t-\chi(t)+\theta)P(\iota,\kappa)x(t-\chi(t)+\theta)$$
$$-(\beta_{1_{\iota\kappa}}+\beta_{2_{\iota\kappa}})\delta x^T(t)P(\iota,\kappa)x(t)\Big]d\theta, \qquad (9.20)$$

where

$$\alpha = min\{\lambda_{min}(-\mathscr{M}_{\iota\kappa}(\hbar(\iota,\kappa),1))\}.$$

It is easy to see that $\alpha > 0$.

Then by Dynkin's formula , we have the following result:

$$
\begin{aligned}
&\mathbf{E}\{V(x(t),\eta_1(t),\eta_2(t),t)\} - \mathbf{E}\{V(x(0),\eta_1(0),\eta_2(0),0)\} \\
&< -\alpha\mathbf{E}\{\int_0^t x^T(s)x(s)ds\} + \int_{-\chi(t)}^{0}\Big[\beta_{1_{\iota\kappa}}\int_0^t \mathbf{E}\{x^T(s+\theta)P(\iota,\kappa)x(s+\theta)\}ds \\
&+\beta_{2_{\iota\kappa}}\int_0^t \mathbf{E}\{x^T(s-\chi(s)+\theta)P(\iota,\kappa)x(s-\chi(s)+\theta)\}ds \\
&-(\beta_{1_{\iota\kappa}}+\beta_{2_{\iota\kappa}})\int_0^t \mathbf{E}\{\delta x^T(s)P(\iota,\kappa)x(s)\}ds\Big]d\theta. \qquad (9.21)
\end{aligned}
$$

Make use of inequality (9.20) and for any $t \geq 0$ and any $x = \{x(\xi) : t - 2\chi(t) \leq \xi \leq t\}$ satisfying (9.8), we have

$$
\mathbf{E}\Big[\max_{\eta_1(t)\in\mathscr{S},\eta_2(t)\in\mathscr{W}} \tilde{A}V(x(t),\eta_1(t),\eta_2(t),t)\Big] \leq -\alpha\mathbf{E}\Big[\|x(t)\|^2\Big]. \qquad (9.22)
$$

Since $\alpha > 0$, following (9.16) we can get

$$
\begin{aligned}
&\mathbf{E}\Big[\max_{\eta_1(t)\in\mathscr{S},\eta_2(t)\in\mathscr{W}} \tilde{A}V(x(t),\eta_1(t),\eta_2(t),t)\Big] \\
&\leq -\frac{\alpha}{\alpha_2}\mathbf{E}\Big[\max_{\eta_1(t)\in\mathscr{S},\eta_2(t)\in\mathscr{W}} V(x(t),\eta_1(t),\eta_2(t),t)\Big].
\end{aligned}
$$

Hence (9.7) is satisfied, which implies that (9.6) is stochastically stable with Markovian jumps. This completes the proof. □

It should be noted that matrix inequalities in Theorem 9.1 are not convex constraints, which are difficult to solve. We therefore propose the following algorithm to change this non-convex feasibility problem into quasi-convex optimization problems [122].

Algorithm 9.1. *ILMI algorithm*

Step 1. Find $Q(\iota,\kappa)$ subject to the following LMI constraints:

$$
\begin{bmatrix} \overline{\Omega}_i(\iota,\kappa) & (*)^T & (*)^T \\ S^T(\iota,\kappa) & -\mathscr{Q}_1 & (*)^T \\ Z^T(\iota,\kappa) & 0 & -\mathscr{Q}_2 \end{bmatrix} < 0, \quad \text{for } i \in \mathscr{I}_R \qquad (9.23)
$$

$$
\begin{bmatrix} \overline{\Omega}_{ij}(\iota,\kappa) & (*)^T & (*)^T \\ S^T(\iota,\kappa) & -\mathscr{Q}_1 & (*)^T \\ Z^T(\iota,\kappa) & 0 & -\mathscr{Q}_2 \end{bmatrix} < 0, \quad \text{for } i < j < r \qquad (9.24)
$$

where

$$\overline{\Omega}_i(\iota,\kappa) = Q(\iota,\kappa)A_i^T + A_iQ(\iota,\kappa) + Y_i^T(\iota,\kappa)B_i^T + B_iY_i(\iota,\kappa),$$
$$\overline{\Omega}_{ij}(\iota,\kappa) = \frac{1}{2}\Big[Q(\iota,\kappa)A_i^T + A_iQ(\iota,\kappa) + Q(\iota,\kappa)A_j^T + A_jQ(\iota,\kappa),$$
$$+Y_j^T(\iota,\kappa)B_i^T + B_iY_j(\iota,\kappa) + Y_i^T(\iota,\kappa)B_j^T + B_jY_i(\iota,\kappa)\Big],$$
$$S(\iota,\kappa) = [\sqrt{\lambda_{\iota 1}}Q(\iota,\kappa)\cdots\sqrt{\lambda_{\iota(\iota-1)}}Q(\iota,\kappa)$$
$$\sqrt{\lambda_{\iota(\iota+1)}}Q(\iota,\kappa)\cdots\sqrt{\lambda_{\iota s}}Q(\iota,\kappa)],$$
$$Z(\iota,\kappa) = [\sqrt{\pi_{\kappa 1}}Q(\iota,\kappa)\cdots\sqrt{\pi_{\kappa(\kappa-1)}}Q(\iota,\kappa)$$
$$\sqrt{\pi_{\kappa(\kappa+1)}}Q(\iota,\kappa)\cdots\sqrt{\pi_{\kappa w}}Q(\iota,\kappa)],$$
$$\mathscr{Q}_1 = diag\{Q(1,\kappa),\cdots,Q(\iota-1,\kappa)\ Q(\iota+1,\kappa),\cdots,Q(s,\kappa)\},$$
$$\mathscr{Q}_2 = diag\{Q(\iota,1),\cdots,Q(\iota,\kappa-1)\ Q(\iota,\kappa+1),\cdots,Q(\iota,w)\}.$$

Let $n = 1$ and $P_n(\iota,\kappa) = Q^{-1}(\iota,\kappa)$. $\varepsilon_{1_{i\iota\kappa}}$, $\varepsilon_{2_{ij\iota\kappa}}$, $\varepsilon_{3_{ij\iota\kappa}}$, and then we set $\varepsilon_{ij_{\iota\kappa}}$ to be 1.

Step 2. Solve the following optimization problem for α_n, $K_i(\iota,\kappa)$, $\beta_{1_{\iota\kappa}}$, $\beta_{2_{\iota\kappa}}$, and $\beta_{3_{\iota\kappa}}$ with the given $P_n(\iota,\kappa)$, $\varepsilon_{1_{i\iota\kappa}}$, $\varepsilon_{2_{ij\iota\kappa}}$, $\varepsilon_{3_{ij\iota\kappa}}$, and $\varepsilon_{ij_{\iota\kappa}}$obtained in the previous step:

OP1 : Minimize α_n subject to the following LMI constraints:

$$\begin{bmatrix} \Omega_i(\iota,\kappa) - \alpha_n P_n(\iota,\kappa) & (*)^T & (*)^T \\ H_i^T P(\iota,\kappa) & -\varepsilon_{ii_{\iota\kappa}} I & (*)^T \\ \varepsilon_{ii_{\iota\kappa}}(E_{1i} + E_{2i}K_i(\iota,\kappa)) & 0 & -\varepsilon_{ii_{\iota\kappa}} I \end{bmatrix} < 0, \quad \text{for } i \in \mathscr{I}_R \tag{9.25}$$

$$\begin{bmatrix} \Omega_{ij}(\iota,\kappa) - \alpha_n P_n(\iota,\kappa) & (*)^T & (*)^T \\ H_i^T P(\iota,\kappa) & -2\varepsilon_{ij_{\iota\kappa}} I & (*)^T \\ \varepsilon_{ij_{\iota\kappa}} \Upsilon_{ij} & 0 & -2\varepsilon_{ij_{\iota\kappa}} I \end{bmatrix} < 0, \quad \text{for } i < j < r \tag{9.26}$$

and (9.11)-(9.12).

Step 3. If $\alpha_n < 0$, $K_i(\iota,\kappa)$ and $P_n(\iota,\kappa)$ are a feasible solution to the BMIs and stop.

Step 4. Set $n = n+1$. Solve the following optimization problem for α_n and $P_n(\iota,\kappa)$, $\varepsilon_{1_{i\iota\kappa}}$, $\varepsilon_{2_{ij\iota\kappa}}$, $\varepsilon_{3_{ij\iota\kappa}}$, and $\varepsilon_{ij_{\iota\kappa}}$ with $K_i(\iota,\kappa)$, $\beta_{1_{\iota\kappa}}$, $\beta_{2_{\iota\kappa}}$, and $\beta_{3_{\iota\kappa}}$ obtained in the previous step:

OP2 : Minimize α_n subject to LMI constraints (9.25), (9.26), and (9.11)-(9.12).

Step 5. If $\alpha_n < 0$, $K_i(\iota,\kappa)$ and $P_n(\iota,\kappa)$ are a feasible solution to the BMIs and stop.

Step 6. Set $n = n+1$. Solve the following optimization problem for $P_n(\iota,\kappa)$, $\varepsilon_{1_{i\iota\kappa}}$, $\varepsilon_{2_{ij\iota\kappa}}$, $\varepsilon_{3_{ij\iota\kappa}}$, and $\varepsilon_{ij_{\iota\kappa}}$ with α_n, $K_i(\iota,\kappa)$, $\beta_{1_{\iota\kappa}}$, $\beta_{2_{\iota\kappa}}$, and $\beta_{3_{\iota\kappa}}$ obtained in the previous step:

OP3 : Minimize trace($P_n(\iota,\kappa)$) subject to LMI constraints (9.25), (9.26), and (9.11)-(9.12).

Step 7. If $\| P_n(\iota,\kappa) - P_{n-1}(\iota,\kappa) \| / \| P_n(\iota,\kappa) \| < \zeta$, ζ is a prescribed tolerance, go to Step 8. Else, set $n = n+1$, $P_n(\iota,\kappa) = P_{n-1}(\iota,\kappa)$, then go to Step 2.

Step 8. A fuzzy state feedback controller for the uncertain nonlineat system may not be found, stop.

Remark 9.1.

1. In Step 1, the initial data is obtained by assuming that the system has no time-delay, that is, $\hbar(\iota,\kappa) = 0$, and uncertainty free, i.e., $\varepsilon_{1_{i\iota\kappa}}$, $\varepsilon_{2_{ij\iota\kappa}}$, $\varepsilon_{3_{ij\iota\kappa}}$, and $\varepsilon_{ij_{\iota\kappa}}$ equal to zero. This guarantees a solution for initial solution for $P_n(\iota,\kappa)$. Furthermore, pre- and post-multiplying (9.9) and (9.10) with $\begin{bmatrix} Q(\iota,\kappa) & 0 & 0 \\ 0 & I & 0 \\ 0 & 0 & I \end{bmatrix}$ results in LMI constraints (9.23) and (9.24).
2. A term $-\alpha_n \begin{bmatrix} P_n(\iota,\kappa) & 0 & 0 \\ 0 & 0 & 0 \\ 0 & 0 & 0 \end{bmatrix}$ is introduced in (9.9) and (9.10) to relax the LMI constraints. It is referred as $\alpha/2$-stabilizable problem in [123]. If an $\alpha_n < 0$ can be found, the fuzzy state feedback controller can be obtained. The rationale behind this concept can also be found in [98].
3. The optimization problem in Step 2 and Step 4 is a generalized eigenvalue minimization problem. These two steps guarantee the progressive reduction of α_n. Step 6 guarantees the convergence of the algorithm.

9.4 Numerical Example

To illustrate the validation of the results obtained in this chapter, we consider a nonlinear mass-spring-damper mechanical system [126] illustrated in Figure 9.2:

$$M\ddot{\theta}(t) + D(\dot{\theta}(t))\dot{\theta}(t) + k\theta(t) = F(t),$$

where $\theta(t)$ is the relative position of the mass; $F(t)$ the external force; $M = 1$ the mass of this system; $k = 0.1$ the stiffness of the spring. The damping coefficient of the nonlinear damper is assumed to be $D(\dot{\theta}(t)) = 0.5 + 0.75\ddot{\theta}(t)$. k is assumed to be bounded within 10% of its nominal value.

Choosing the stats as $x_1(t) = \dot{\theta}(t)$ and $x_2(t) = \theta(t)$ and the input variable $u(t)$ as $F(t)$ yields the following state space representation:

$$\begin{bmatrix} \dot{x}_1(t) \\ \dot{x}_2(t) \end{bmatrix} = \begin{bmatrix} -0.75x_1^3(t) - 0.5x_1(t) + 0.1(1+0.1F(t))x_2(t) + u(t) \\ x_1(t) \end{bmatrix}.$$

Assume $x_1(t) \in [-1,1]$, we address the nonlinear term $-0.75x_1^3(t)$ using the methodology given in [124].

The T-S fuzzy system is therefore constructed as follows:

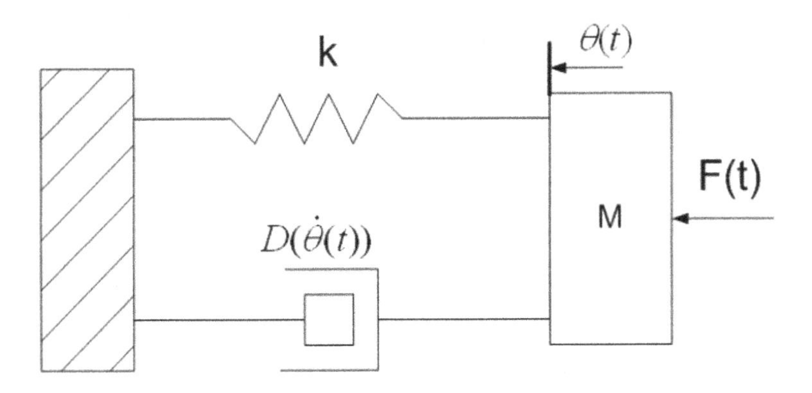

Fig. 9.2 A nonlinear mass-spring-damper system

Plant Rule 1:
IF $x_1(t)$ is about N_1,
THEN $\dot{x}(t) = (A_1 + \Delta A_1)x(t) + (B_1 + \Delta B_1)u(t)$.
Plant Rule 2:
IF $x_1(t)$ is about N_2,
THEN $\dot{x}(t) = (A_2 + \Delta A_2)x(t) + (B_2 + \Delta B_2)u(t)$.
where

$$N_1(x_1(t)) = 1 - x_1^2(t), N_2(x_1(t)) = x_1^2(t).$$

$$A_1 = \begin{bmatrix} -0.5 & 0.1 \\ 1 & 0 \end{bmatrix}, A_2 = \begin{bmatrix} -1 & 0.1 \\ 1 & 0 \end{bmatrix},$$

$$\Delta A_1 = \Delta A_2 = \begin{bmatrix} 0 & 0.1F(t) \\ 0 & 0 \end{bmatrix},$$

$$B_1 = B_2 = \begin{bmatrix} 1 \\ 0 \end{bmatrix}, \Delta B_1 = \Delta B_2 = \begin{bmatrix} 0 \\ 0 \end{bmatrix}.$$

In the following simulation, we assume $F(t) = \sin t$, therefore it satisfies Assumption 9.1.

Furthermore, we assume that the sensor-to-controller communication delays for two Markovian modes are $|\tau_1| < 0.02$, $|\tau_2| < 0.015$, while the controller-to-actuator delays are $|\rho_1| < 0.02$, and $|\rho_2| < 0.015$, and therefore we can have $\hbar_{11} = 0.04$, $\hbar_{12} = 0.035$, $\hbar_{21} = 0.035$, and $\hbar_{22} = 0.03$. We assume that the random time-delays exist in $\mathscr{S} = \{1,2\}$ and $\mathscr{W} = \{1,2\}$, and their transition rate matrices are given by:

$$\Lambda = \begin{bmatrix} -3 & 3 \\ 2 & -2 \end{bmatrix}, \Pi = \begin{bmatrix} -1 & 1 \\ 2 & -2 \end{bmatrix}.$$

A controller of the form (9.4) is obtained using Theorem 9.1 and the algorithm stated in the previous section, that is,

$$K_1(1,1) = [-0.3615 - 1.0192], K_1(1,2) = [-0.7553 - 1.1221],$$
$$K_1(2,1) = [-0.2452 - 0.8808], K_1(2,2) = [-0.5533 - 1.0503],$$
$$K_2(1,1) = [-0.2590 - 1.0442], K_2(1,2) = [-0.6360 - 1.1464],$$
$$K_2(2,1) = [-0.1172 - 0.8919], K_2(2,2) = [-0.4256 - 1.0809].$$

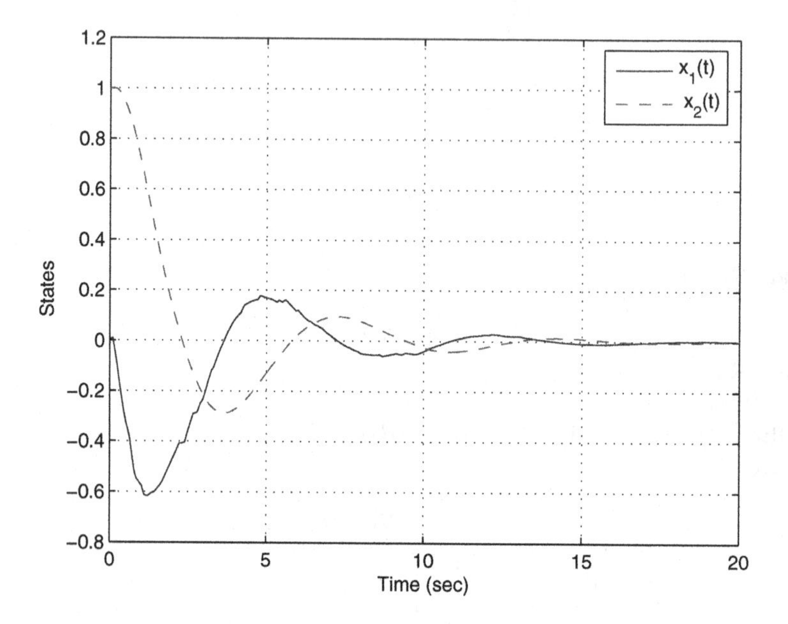

Fig. 9.3 Response of plant states

Remark 9.2. The state trajectories of the closed-loop system are shown in Figure 9.3 with initial states chosen as $x(0) = x_0 = [0\ 1]^T$. It can be seen that the system is stochastically stable, which demonstrates the validity of the methodology put forward in this chapter.

9.5 Conclusion

In this chapter, a technique of designing a mode-dependent state feedback controller for a nonlinear uncertain NCSs with communication random time-delays has been proposed. The design process is based on a T-S fuzzy model that approximates the nonlinear system. The main contribution of this work is that both the sensor-to-controller and controller-to-actuator delays have been taken into account. Two

Markov processes have been used to model these two time-delays. The Lyapunov–Razumikhin method has been employed to derive a mode dependent state feedback for this class of systems. Sufficient conditions for the existence of a mode-dependent state feedback controller for this class of NCSs are derived. We use a numerical example to demonstrate the effectiveness of this methodology at the last section.

Chapter 10
Dynamic Output Feedback Control for Uncertain Nonlinear Networked Control Systems

In this chapter, a fuzzy dynamic output feedback controller is designed for a class of uncertain nonlinear NCSs. The design procedure is inherited from the previous chapter. The results are given in terms of the solvability of BMIs.

10.1 Problem Formulation and Preliminaries

A class of uncertain nonlinear systems under consideration in this chapter is described by the following IF-THEN rules and the ith rule has been shown as follows:

Plant Rule i:
IF $\nu_1(t)$ is M_{i1} and $\cdots$ and $\nu_p(t)$ is M_{ip},
THEN

$$\begin{cases} \dot{x}(t) = (A_i + \Delta A_i)x(t) + (B_i + \Delta B_i)u(t) \\ y(t) = (C_i + \Delta C_i)x(t) \end{cases} \tag{10.1}$$

where $i \in \mathscr{I}_R = \{1, \cdots, r\}$, r is the number of fuzzy rules; $\nu_k(t)$ are premise variables, $M_{\iota\kappa}$ are fuzzy sets, $k = 1, \cdots, p$, p is the number of premise variables; $x(t) \in \Re^n$ and $u(t) \in \Re^m$ denote state and control input, respectively. Matrices $A_i \in \Re^{n\times n}$ and $B_i \in \Re^{n\times m}$ are known system matrices. Matrices ΔA_i and ΔB_i represent the uncertainties in the system and satisfy the following assumption.

Assumption 10.1. The parameter uncertainties considered here are norm-bounded, in the form

$$\begin{aligned} \left[\Delta A_i \ \Delta B_i \right] &= H_{1i} F_i(t) \left[E_{1i} \ E_{2i} \right], \\ \Delta C_i &= H_{2i} F_i(t) E_{1i}, \end{aligned}$$

where H_{1i}, H_{2i}, E_{1i} and E_{2i} are known real constant matrices of appropriate dimensions, and $F_i(t)$ is an unknown matrix function with Lebesgue-measurable elements and satisfies $F_i(t)^T F_i(t) \leq I$, in which I is the identity matrix of appropriate dimension.

D. Huang and S.K. Nguang: Robust Ctrl. for Uncertain Networked Ctrl. Sys., LNCIS 386, pp. 107–115.
springerlink.com

By using a center-average defuzzifier, product inference and singleton fuzzifier, the local models can be integrated into a global nonlinear model:

$$\begin{cases} \dot{x}(t) = \sum_{i=1}^{r} \mu_i(v(t))[(A_i + \Delta A_i)x(t) + (B_i + \Delta B_i)u(t)] \\ y(t) = \sum_{i=1}^{r} \mu_i(v(t))(C_i + \Delta C_i)x(t) \end{cases} \tag{10.2}$$

where

$$v(t) = [v_1(t), v_2(t), \cdots, v_p(t)]^T,$$

$$\omega_i(v(t)) = \prod_{k=1}^{p} M_{\iota\kappa}(v_k(t)),\ \omega_i(v(t)) \geq 0,\ \sum_{i=1}^{r} \omega_i(v(t)) > 0,$$

$$\mu_i(v(t)) = \frac{\omega_i(v(t))}{\sum_{i=1}^{r} \omega_i(v(t))},\ \mu_i(v(t)) \geq 0,\ \sum_{i=1}^{r} \mu_i(v(t)) = 1.$$

Here, $M_{\iota\kappa}(v_k(t))$ denote the grade of membership of $v_k(t)$ in $M_{\iota\kappa}$.

In this chapter, we consider a nonlinear NCS of which the plant is described by the T-S model (10.2). The setup of the overall control system is depicted in Figure 10.1, where $\tau(t) \geq 0$ is the random time-delay from sensor to controller and $\rho(t) \geq 0$ is the random time-delay from controller to actuator. These delays are assumed to be upper bounded. We apply the same Markov processes introduced in Chapter 2 to model the random time-delays in this chapter.

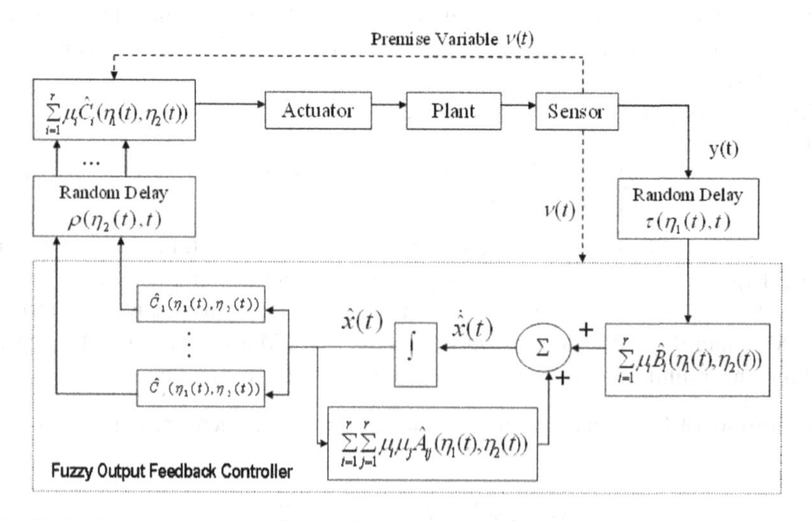

Fig. 10.1 Block diagram of a nonlinear NCS with a dynamic output feedback controller

In the system setup, the premise vector $v(t)$ is connected to the controller and actuators via point-to-point architecture, which is immune to network-induced delays.

Therefore, following the modeling procedure presented in Chapter 2, for the non-linear plant represented by (10.2), the fuzzy dynamic output feedback controller at time t is inferred as follows:

$$\begin{cases} \dot{\hat{x}}(t) = \sum_{i=1}^{r}\sum_{j=1}^{r}\mu_i(v(t))\mu_j(v(t))\Big[\hat{A}_{ij}(\eta_1(t),\eta_2(t))\hat{x}(t) \\ \qquad\qquad +\hat{B}_i(\eta_1(t),\eta_2(t))y(t-\tau(\eta_1(t),t))\Big] \\ u(t) = \sum_{i=1}^{r}\mu_i(v(t))\hat{C}_i(\eta_1(t),\eta_2(t)\hat{x}(t) \end{cases} \tag{10.3}$$

where $\hat{A}_{ij}(\eta_1(t),\eta_2(t))$, $\hat{B}_i(\eta_1(t),\eta_2(t))$, $\hat{C}_i(\eta_1(t),\eta_2(t))$ in each plant rule are parameters of the controller which are to be designed. The plant model is rewritten as:

$$\begin{cases} \dot{x}(t) = \sum_{i=1}^{r}\mu_i(v(t))[(A_i+\Delta A_i)x(t)+(B_i+\Delta B_i)u(t-\rho(\eta_2(t),t))] \\ y(t) = \sum_{i=1}^{r}\mu_i(v(t))(C_i+\Delta C_i)x(t) \end{cases} \tag{10.4}$$

Substituting (10.3) into (10.4) yields

$$\begin{aligned} \dot{\tilde{x}}(t) = {} & \sum_{i=1}^{r}\sum_{j=1}^{r}\mu_i(v(t))\mu_j(v(t))\Big[\tilde{A}_{ij}(\eta_1(t),\eta_2(t))\tilde{x}(t) \\ & +\tilde{B}_{ij}(\eta_1(t),\eta_2(t))\tilde{x}(t-\tau(\eta_1(t),t))+\tilde{C}_{ij}(\eta_1(t),\eta_2(t))\tilde{x}(t-\rho(\eta_1(t),t))\Big], \end{aligned} \tag{10.5}$$

where

$$\tilde{x}(t) = \begin{bmatrix} x(t) \\ \hat{x}(t) \end{bmatrix},\ \tilde{A}_{ij}(\eta_1(t),\eta_2(t)) = \begin{bmatrix} A_i+\Delta A_i & 0 \\ 0 & \hat{A}_{ij}(\eta_1(t),\eta_2(t)) \end{bmatrix},$$
$$\tilde{B}_{ij}(\eta_1(t),\eta_2(t)) = \begin{bmatrix} 0 & 0 \\ \hat{B}_i(\eta_1(t),\eta_2(t))(C_j+\Delta C_j) & 0 \end{bmatrix},$$
$$\tilde{C}_{ij}(\eta_1(t),\eta_2(t)) = \begin{bmatrix} 0 & (B_i+\Delta B_i)\hat{C}_j(\eta_1(t),\eta_2(t)) \\ 0 & 0 \end{bmatrix}.$$

The aim of this chapter is to design a dynamic output feedback controller of the form (10.3) such that the following inequality holds:

$$\begin{aligned} & \mathbf{E}\Big[\max_{\eta_1(t)\in\mathscr{S},\eta_2(t)\in\mathscr{W}}\tilde{A}V(x(t),\eta_1(t),\eta_2(t),t)\Big] \\ & \le -\zeta\mathbf{E}\Big[\max_{\eta_1(t)\in\mathscr{S},\eta_2(t)\in\mathscr{W}}V(x(t),\eta_1(t),\eta_2(t),t)\Big], \end{aligned} \tag{10.6}$$

provided $x = \{x(\xi): t-2\chi \le \xi \le t\}$ satisfying:

$$\begin{aligned} & \mathbf{E}\Big[\min_{\eta_1(t)\in\mathscr{S},\eta_2(t)\in\mathscr{W}}V(x(\xi),\eta_1(\xi),\eta_2(\xi),\xi)\Big] \\ & < \delta\mathbf{E}\Big[\max_{\eta_1(t)\in\mathscr{S},\eta_2(t)\in\mathscr{W}}V(x(t),\eta_1(t),\eta_2(t),t)\Big], \end{aligned} \tag{10.7}$$

for all $t-2\chi \leq \xi \leq t$. Here $\tilde{A}$ denotes the weak infinitesimal operator and $\mathbf{E}[\cdot]$ stands for the mathematical expectation.

Then the system (10.5) is said to achieve stochastic stability with Markovian jumps.

In this chapter, we assume $u(t) = 0$ before the first control signal reaches the plant. Moreover, $\mu_i(\nu(t))$ and $\mu_j(\nu(t))$ are denoted as μ_i and μ_j respectively for the convenience of notations. In the symmetric block matrices, we use (*) as an ellipsis for terms that are induced by symmetry. $\hat{A}_{ij}(\eta_1(t), \eta_2(t))$ is denoted as $\hat{A}_{ij}(\iota, \kappa)$ if $\eta_1(t) = \iota$ and $\eta_2(t) = \kappa$.

10.2 Main Result

The following theorem provides sufficient conditions for the existence of a mode-dependent dynamic output feedback controller for the system (10.5).

Theorem 10.1. *Consider the system (10.5) satisfying Assumption 10.1. Given positive scalars $\hbar(\iota, \kappa)$, $\varepsilon_{1ij_{\iota\kappa}}$, $\varepsilon_{2ij_{\iota\kappa}}$, $\varepsilon_{3ij_{\iota\kappa}}$, $\varepsilon_{4ij_{\iota\kappa}}$, and $\varepsilon_{5ij_{\iota\kappa}}$, if there exist symmetric matrices $X(\iota, \kappa)$, $Y(\iota, \kappa)$, matrices $L_i(\iota, \kappa)$, $F_i(\iota, \kappa)$, and positive scalars $\beta_{1_{\iota\kappa}}$, $\beta_{2_{\iota\kappa}}$, such that the following inequalities hold for all $\iota \in \mathscr{S}$ and $\kappa \in \mathscr{W}$:*

$$\begin{bmatrix} Y(\iota,\kappa) & I \\ I & X(\iota,\kappa) \end{bmatrix} > 0, \tag{10.8}$$

$$\Omega_{ii}(\iota,\kappa) < 0, \text{ for } i \in \mathscr{I}_R \tag{10.9}$$

$$\Omega_{ij}(\iota,\kappa) + \Omega_{ji}(\iota,\kappa) < 0, \text{ for } i < j < r \tag{10.10}$$

$$\Upsilon_{ij}(\iota,\kappa) < 0, \text{ for } \{i,j\} \in \mathscr{I}_R \times \mathscr{I}_R \tag{10.11}$$

$$\begin{bmatrix} R_{4_{\iota\kappa}} & (*)^T \\ \Lambda_\iota^T & \mathscr{Q}_{1_{\iota\kappa}} \end{bmatrix} > 0, \tag{10.12}$$

$$\begin{bmatrix} R_{5_{\iota\kappa}} & (*)^T \\ \Pi_\kappa^T & \mathscr{Q}_{2_{\iota\kappa}} \end{bmatrix} > 0, \tag{10.13}$$

$$\begin{bmatrix} -R_{1_{\iota\kappa}} & (*)^T & (*)^T & (*)^T \\ 0 & -I & (*)^T & (*)^T \\ 0 & -Y(\iota,\kappa) & -R_{2_{\iota\kappa}} & (*)^T \\ 0 & 0 & 0 & -R_{3_{\iota\kappa}} \end{bmatrix} < 0, \tag{10.14}$$

$$\begin{bmatrix} -\beta_{2_{\iota\kappa}}Y(\iota,\kappa) & (*)^T & (*)^T & (*)^T & (*)^T & (*)^T \\ -\beta_{2_{\iota\kappa}}I & -\beta_{2_{\iota\kappa}}X(\iota,\kappa) & (*)^T & (*)^T & (*)^T & (*)^T \\ L_j^T(\iota,\kappa)B_i^T & L_j^T(\iota,\kappa)B_i^T X(\iota,\kappa) & -Y(\iota,\kappa) & (*)^T & (*)^T & (*)^T \\ 0 & 0 & -I & -X(\iota,\kappa) & (*)^T & (*)^T \\ \varepsilon_{4ij_{\iota\kappa}}H_{1i}^T & \varepsilon_{4ij_{\iota\kappa}}H_{1i}^T X(\iota,\kappa) & 0 & 0 & -\varepsilon_{4ij_{\iota\kappa}}I & (*)^T \\ 0 & 0 & E_{2i}L_j(\iota,\kappa) & 0 & 0 & -\varepsilon_{4ij_{\iota\kappa}}I \end{bmatrix} < 0,$$

$$\text{for } \{i,j\} \in \mathscr{I}_R \times \mathscr{I}_R \tag{10.15}$$

$$\begin{bmatrix} -\beta_{2_{\iota\kappa}}Y(\iota,\kappa) & (*)^T & (*)^T & (*)^T & (*)^T & (*)^T \\ -\beta_{2_{\iota\kappa}}I & -\beta_{2_{\iota\kappa}}X(\iota,\kappa) & (*)^T & (*)^T & (*)^T & (*)^T \\ 0 & Y(\iota,\kappa)C_j^T F_i^T(\iota,\kappa) & -Y(\iota,\kappa) & (*)^T & (*)^T & (*)^T \\ 0 & 0 & -I & -X(\iota,\kappa) & (*)^T & (*)^T \\ 0 & \varepsilon_{5ij_{\iota\kappa}}H_{2j}^T F_i^T(\iota,\kappa) & 0 & 0 & -\varepsilon_{5ij_{\iota\kappa}}I & (*)^T \\ 0 & 0 & E_{1j}Y(\iota,\kappa) & 0 & 0 & -\varepsilon_{5ij_{\iota\kappa}}I \end{bmatrix} < 0,$$

$$\text{for } \{i,j\} \in \mathscr{I}_R \times \mathscr{I}_R \tag{10.16}$$

where

$$\Omega_{ij}(\iota,\kappa) = \begin{bmatrix} \begin{pmatrix} A_iY(\iota,\kappa)+Y(\iota,\kappa)A_i^T \\ +B_iL_j(\iota,\kappa)+L_j^T(\iota,\kappa)B_i^T \\ +(\beta_{1_{\iota\kappa}}+5\beta_{2_{\iota\kappa}})\hbar_{\iota\kappa}Y(\iota,\kappa) \\ +(\lambda_{\iota\iota}+\pi_{\kappa\kappa})Y(\iota,\kappa) \end{pmatrix} & (*)^T & (*)^T & (*)^T & (*)^T & (*)^T & (*)^T & (*)^T \\ (\beta_{1_{\iota\kappa}}+5\beta_{2_{\iota\kappa}})\hbar_{\iota\kappa}I & \begin{pmatrix} X(\iota,\kappa)A_i+A_i^TX(\iota,\kappa) \\ +F_i(\iota,\kappa)C_j+C_j^TF_i^T(\iota,\kappa) \\ +(\beta_{1_{\iota\kappa}}+5\beta_{2_{\iota\kappa}})\hbar_{\iota\kappa}X(\iota,\kappa) \\ +\sum_{\wp=1}^{s}\lambda_{\iota\wp}X(\wp,\kappa) \\ +\sum_{\ell=1}^{w}\pi_{\kappa\ell}X(\iota,\ell) \end{pmatrix} & (*)^T & (*)^T & (*)^T & (*)^T & (*)^T & (*)^T \\ \varepsilon_{1ij_{\iota\kappa}}H_{1i}^T & \varepsilon_{1ij_{\iota\kappa}}H_{1i}^TX(\iota,\kappa) & -\varepsilon_{1ij_{\iota\kappa}}I & 0 & (*)^T & (*)^T & (*)^T & (*)^T \\ E_{1i}Y(\iota,\kappa)+E_{2i}L_j(\iota,\kappa) & E_{1i} & 0 & -\varepsilon_{1ij_{\iota\kappa}}I & (*)^T & (*)^T & (*)^T & (*)^T \\ 0 & \varepsilon_{2ij_{\iota\kappa}}H_{2j}^TF_i^T(\iota,\kappa) & 0 & 0 & -\varepsilon_{2ij_{\iota\kappa}}I & (*)^T & (*)^T & (*)^T \\ E_{1j}Y(\iota,\kappa) & E_{1j} & 0 & 0 & 0 & -\varepsilon_{2ij_{\iota\kappa}}I & (*)^T & (*)^T \\ S^T(\iota,\kappa) & 0 & 0 & 0 & 0 & 0 & -\mathscr{Q}_{1_{\iota\kappa}} & (*)^T \\ Z^T(\iota,\kappa) & 0 & 0 & 0 & 0 & 0 & 0 & -\mathscr{Q}_{2_{\iota\kappa}} \end{bmatrix}$$

$$\Upsilon_{ij}(\iota,\kappa)=\begin{bmatrix} -\beta_{1_{\iota\kappa}}Y(\iota,\kappa)+2R_{1_{\iota\kappa}} & (*)^T & (*)^T \\ -\beta_{1_{\iota\kappa}}I & -\beta_{1_{\iota\kappa}}X(\iota,\kappa) & (*)^T \\ Y(\iota,\kappa)A_i^T & \begin{pmatrix} -A_i-L_j^T(\iota,\kappa)B_i^TX(\iota,\kappa) \\ -Y(\iota,\kappa)C_j^TF_i^T(\iota,\kappa) \\ -(\lambda_{\iota\iota}+\pi_{\kappa\kappa})I \end{pmatrix} & -Y(\iota,\kappa)+2R_{2_{\iota\kappa}} \\ A_i^T & A_i^TX(\iota,\kappa) & -I \\ 0 & R_{4_{\iota\kappa}} & 0 \\ 0 & R_{5_{\iota\kappa}} & 0 \\ \varepsilon_{3ij_{\iota\kappa}}H_{1i}^T & \varepsilon_{3ij_{\iota\kappa}}H_{1i}^TX(\iota,\kappa) & 0 \\ 0 & 0 & E_{1i}Y(\iota,\kappa) \end{bmatrix.}$$

$$\begin{matrix} (*)^T & (*)^T & (*)^T & (*)^T & (*)^T \\ (*)^T & (*)^T & (*)^T & (*)^T & (*)^T \\ (*)^T & (*)^T & (*)^T & (*)^T & (*)^T \\ -X(\iota,\kappa)+2R_{3_{\iota\kappa}} & (*)^T & (*)^T & (*)^T & (*)^T \\ 0 & -I & (*)^T & (*)^T & (*)^T \\ 0 & 0 & -I & (*)^T & (*)^T \\ 0 & 0 & 0 & -\varepsilon_{3ij_{\iota\kappa}}I & (*)^T \\ E_{1i} & 0 & 0 & 0 & -\varepsilon_{3ij_{\iota\kappa}}I \end{matrix}\Bigg]$$

and

$$\begin{aligned}
&S(\iota,\kappa)=[\sqrt{\lambda_{\iota 1}}Y(\iota,\kappa)\cdots\sqrt{\lambda_{\iota(\iota-1)}}Y(\iota,\kappa)\ \sqrt{\lambda_{\iota(\iota+1)}}Y(\iota,\kappa)\cdots\sqrt{\lambda_{\iota s}}Y(\iota,\kappa)],\\
&Z(\iota,\kappa)=[\sqrt{\pi_{\kappa 1}}Y(\iota,\kappa)\cdots\sqrt{\pi_{\kappa(\kappa-1)}}Y(\iota,\kappa)\ \sqrt{\pi_{\kappa(\kappa+1)}}Y(\iota,\kappa)\cdots\sqrt{\pi_{\kappa w}}Y(\iota,\kappa)],\\
&\Lambda_\iota=[\sqrt{\lambda_{\iota 1}}I\cdots\sqrt{\lambda_{\iota(\iota-1)}}I\ \sqrt{\lambda_{\iota(\iota+1)}}I\cdots\sqrt{\lambda_{\iota s}}I],\\
&\Pi_\kappa=[\sqrt{\pi_{\kappa 1}}I\cdots\sqrt{\pi_{\kappa(\kappa-1)}}I\ \sqrt{\pi_{\kappa(\kappa+1)}}I\cdots\sqrt{\pi_{\kappa w}}I],\\
&\mathcal{Q}_{1_{\iota\kappa}}=diag\{Y(1,\kappa),\cdots,Y(\iota-1,\kappa),Y(\iota+1,\kappa),\cdots,Y(s,\kappa)\},\\
&\mathcal{Q}_{2_{\iota\kappa}}=diag\{Y(\iota,1),\cdots,Y(\iota,\kappa-1),Y(\iota,\kappa+1),\cdots,Y(\iota,w)\},
\end{aligned}$$

then the system (10.5) is said to achieve stochastic stability via controller (10.3) for all delays $\tau(\iota,t)$ and $\rho(\kappa,t)$ satisfying

$$0\le\tau(\iota,t)+\rho(\kappa,t)\le\hbar(\iota,\kappa).$$

Furthermore, the mode dependant controller is of the form (10.3) with

$$\begin{aligned}
\hat{A}_{ij}(\iota,\kappa)&=[Y^{-1}(\iota,\kappa)-X(\iota,\kappa)]^{-1}[-A_i^T-X(\iota,\kappa)A_iY(\iota,\kappa)-F_i(\iota,\kappa)C_jY(\iota,\kappa)\\
&\quad -X(\iota,\kappa)B_iL_j(\iota,\kappa)-\sum_{\wp=1}^{s}\lambda_{\iota\wp}Y^{-1}(\wp,\kappa)Y(\iota,\kappa)\\
&\quad -\sum_{\ell=1}^{w}\pi_{\kappa\ell}Y^{-1}(\iota,\ell)Y(\iota,\kappa)]Y^{-1}(\iota,\kappa), &(10.17)\\
\hat{B}_i(\iota,\kappa)&=[Y^{-1}(\iota,\kappa)-X(\iota,\kappa)]^{-1}F_i(\iota,\kappa), &(10.18)\\
\hat{C}_i(\iota,\kappa)&=L_i(\iota,\kappa)Y^{-1}(\iota,\kappa). &(10.19)
\end{aligned}$$

Proof. It is straightforward from the proof process in Chapter 9 and Chapter 4. □

The iterative algorithm presented in Chapter 9 is applied here to solve Theorem 10.1 which is a BMI problem.

10.3 Numerical Example

To illustrate the validation of the results obtained in this chapter, we consider the same plant as in the previous chapter. The fuzzy rules and the transition rate matrices remain unchanged.

The maximal random time-delays are bounded as $\hbar(1,1) = 0.05$, $\hbar(1,2) = 0.09$, $\hbar(2,1) = 0.04$, $\hbar(2,2) = 0.07$.

A controller of the form (10.3) is obtained using Theorem 10.1 and the algorithm stated in the previous section, that is,

$$\hat{A}_{11}(1,1) = \begin{bmatrix} -3.3615 & -5.6192 \\ 27.5443 & 44.6709 \end{bmatrix}, \hat{A}_{12}(1,1) = \begin{bmatrix} -3.3611 & -5.6193 \\ 27.5430 & 44.6722 \end{bmatrix},$$

$$\hat{A}_{21}(1,1) = \begin{bmatrix} -3.4783 & -5.5542 \\ 29.0769 & 46.0009 \end{bmatrix}, \hat{A}_{22}(1,1) = \begin{bmatrix} -3.4632 & -5.5465 \\ 29.3422 & 46.9807 \end{bmatrix},$$

$$\hat{B}_1(1,1) = \begin{bmatrix} 0.3627 \\ -1.2254 \end{bmatrix}, \hat{B}_2(1,1) = \begin{bmatrix} 0.4077 \\ -1.0145 \end{bmatrix},$$

$$\hat{C}_1(1,1) = \begin{bmatrix} 63.7888 & 2.7765 \end{bmatrix}, \hat{C}_2(1,1) = \begin{bmatrix} 41.0123 & 2.6663 \end{bmatrix},$$

$$\hat{A}_{11}(1,2) = \begin{bmatrix} -2.6749 & -2.4465 \\ 6.5856 & 9.5678 \end{bmatrix}, \hat{A}_{12}(1,2) = \begin{bmatrix} -2.6666 & -2.4466 \\ 6.5356 & 9.6074 \end{bmatrix},$$

$$\hat{A}_{21}(1,2) = \begin{bmatrix} -2.9776 & -3.0123 \\ 6.2654 & 9.1298 \end{bmatrix}, \hat{A}_{22}(1,2) = \begin{bmatrix} -2.9804 & -3.0007 \\ 6.2709 & 9.1112 \end{bmatrix},$$

$$\hat{B}_1(1,2) = \begin{bmatrix} 0.7733 \\ -1.9998 \end{bmatrix}, \hat{B}_2(1,2) = \begin{bmatrix} 0.5443 \\ -2.0987 \end{bmatrix},$$

$$\hat{C}_1(1,2) = \begin{bmatrix} 41.0987 & 1.9192 \end{bmatrix}, \hat{C}_2(1,2) = \begin{bmatrix} 22.7895 & 3.3345 \end{bmatrix},$$

$$\hat{A}_{11}(2,1) = \begin{bmatrix} -4.3333 & -1.9192 \\ 15.8767 & 23.0962 \end{bmatrix}, \hat{A}_{12}(2,1) = \begin{bmatrix} -4.2976 & -1.9111 \\ 15.8098 & 23.1999 \end{bmatrix},$$

$$\hat{A}_{21}(2,1) = \begin{bmatrix} -5.7620 & -2.1857 \\ 19.7600 & 22.0911 \end{bmatrix}, \hat{A}_{22}(2,1) = \begin{bmatrix} -5.7557 & -2.1770 \\ 19.7858 & 22.1887 \end{bmatrix},$$

$$\hat{B}_1(2,1) = \begin{bmatrix} 0.4333 \\ -2.0098 \end{bmatrix}, \hat{B}_2(2,1) = \begin{bmatrix} 2.9833 \\ -1.0192 \end{bmatrix},$$

$$\hat{C}_1(2,1) = \begin{bmatrix} 31.2314 & 6.9856 \end{bmatrix}, \hat{C}_2(2,1) = \begin{bmatrix} 48.0009 & 0.8876 \end{bmatrix},$$

$$\hat{A}_{11}(2,2) = \begin{bmatrix} -0.1762 & -0.5543 \\ 1.4466 & 3.2009 \end{bmatrix}, \hat{A}_{12}(2,2) = \begin{bmatrix} -0.1762 & -0.5543 \\ 1.4466 & 3.2009 \end{bmatrix},$$

$$\hat{A}_{21}(2,2) = \begin{bmatrix} -0.3033 & -0.4031 \\ 2.5499 & 4.0988 \end{bmatrix}, \hat{A}_{22}(2,2) = \begin{bmatrix} -0.3111 & -0.4232 \\ 2.6778 & 4.2212 \end{bmatrix},$$

$$\hat{B}_1(2,2) = \begin{bmatrix} 1.2331 \\ -5.2312 \end{bmatrix}, \hat{B}_2(2,2) = \begin{bmatrix} 2.0933 \\ -7.0212 \end{bmatrix},$$

$$\hat{C}_1(2,2) = \begin{bmatrix} 101.0214 & 7.9856 \end{bmatrix}, \hat{C}_2(2,2) = \begin{bmatrix} 55.8373 & 3.9883 \end{bmatrix}.$$

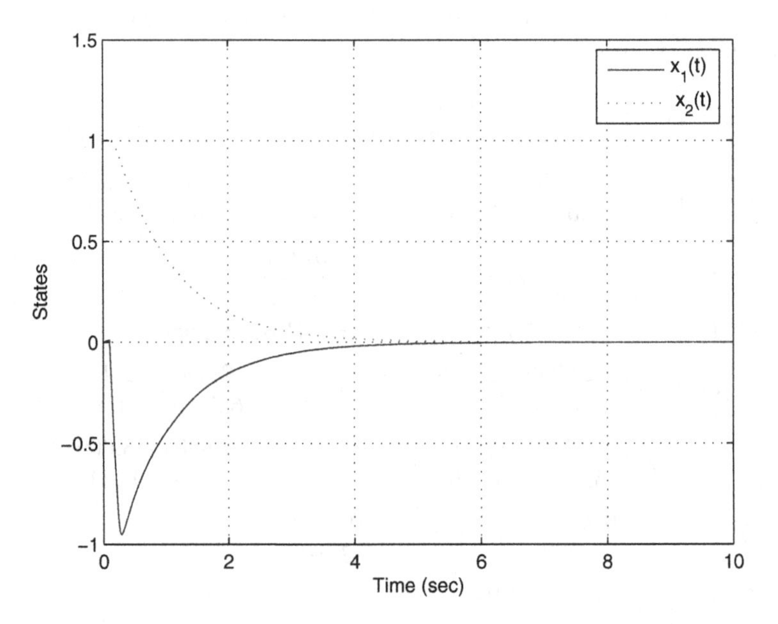

Fig. 10.2 Response of plant states

Remark 10.1. In the simulation, we select $F(t) = \sin(10t)$, $\rho(1,t) = 0.015\sin(10t)$, $\tau(1,t) = \tau(2,t) = 0.03\sin(10t)$, and $\rho(2,t) = 0.03\sin(10t)$. The state trajectories of the closed-loop system are shown in Figure 10.2 with initial states chosen as $x(0) = x_0 = [0\ 1]^T$. The input signal is shown in Figure 10.3. It can be seen that the system is stochastically stable, which demonstrates the validity of the methodology put forward in this chapter.

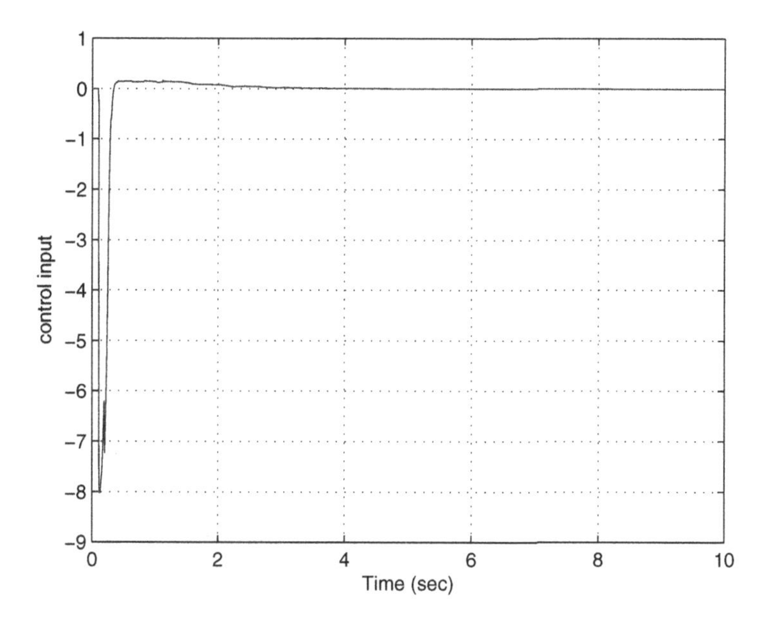

Fig. 10.3 Control input

10.4 Conclusion

In this chapter, a mode-dependent dynamic output feedback controller for nonlinear uncertain NCSs is considered. The Lyapunov–Razumikhin method has been employed to derive a mode-dependent controller for this class of systems. Sufficient conditions for the existence of such a controller for this class of NCSs are derived. We use a numerical example to demonstrate the effectiveness of this methodology in the last section.

Chapter 11
Robust Disturbance Attenuation for Uncertain Nonlinear Networked Control Systems

In this chapter, we consider the problem of robust fuzzy disturbance attenuation for a class of uncertain nonlinear NCSs. The Lyapunov–Razumikhin method has been employed to derive such a controller for this class of systems such that it is stochastically stabilizable with a disturbance attenuation level γ. Sufficient conditions for the existence of such a controller for this class of NCSs are derived in terms of the solvability of BMIs.

11.1 Problem Formulation and Preliminaries

In this chapter, we describe the nonlinear NCSs as follows:

$$\begin{cases} \dot{x}(t) = \sum_{i=1}^{r} \mu_i(\nu(t))[(A_i + \Delta A_i)x(t) + (B_{1i} + \Delta B_{1i})w(t) + (B_{2i} + \Delta B_{2i})u(t)] \\ z(t) = \sum_{i=1}^{r} \mu_i(\nu(t))[(C_{1i} + \Delta C_{1i})x(t) + (D_{1i} + \Delta D_{1i})u(t)] \\ y(t) = \sum_{i=1}^{r} \mu_i(\nu(t))[(C_{2i} + \Delta C_{2i})x(t) + (D_{2i} + \Delta D_{2i})w(t)] \end{cases} \tag{11.1}$$

where $x(t) \in \mathbb{R}^n$ is the state vector, $u(t) \in \mathbb{R}^m$ is the control input, $w(t) \in \mathbb{R}^p$ is the exogenous disturbance input and/or measurement noise, $y(t) \in \mathbb{R}^l$ and $z(t) \in \mathbb{R}^s$ denote the measurement and regulated output respectively.

Furthermore, $i \in \mathscr{I}_R = \{1, \cdots, r\}$, r is the number of fuzzy rules; $\nu_k(t)$ are premise variables, $M_{\iota\kappa}$ are fuzzy sets, $k = 1, \cdots, p$, p is the number of premise variables

$$\nu(t) = [\nu_1(t), \nu_2(t), \cdots, \nu_p(t)]^T,$$

$$\omega_i(\nu(t)) = \prod_{k=1}^{p} M_{\iota\kappa}(\nu_k(t)),\ \omega_i(\nu(t)) \geq 0,\ \sum_{i=1}^{r} \omega_i(\nu(t)) > 0,$$

$$\mu_i(\nu(t)) = \frac{\omega_i(\nu(t))}{\sum_{i=1}^{r} \omega_i(\nu(t))},\ \mu_i(\nu(t)) \geq 0,\ \sum_{i=1}^{r} \mu_i(\nu(t)) = 1.$$

Here, $M_{\iota\kappa}(\nu_k(t))$ denote the grade of membership of $\nu_k(t)$ in $M_{\iota\kappa}$.

D. Huang and S.K. Nguang: Robust Ctrl. for Uncertain Networked Ctrl. Sys., LNCIS 386, pp. 117–127.
springerlink.com

In addition, matrices ΔA_i, ΔB_{1i}, ΔB_{2i}, ΔC_{1i}, ΔC_{2i}, ΔD_{1i}, and ΔD_{2i} characterize the uncertainties in the system and satisfy the following assumption:

Assumption 11.1.

$$\begin{aligned}\begin{bmatrix}\Delta A_i \ \Delta B_{1i} \ \Delta B_{2i}\end{bmatrix} &= H_{1i}F(t)\begin{bmatrix}E_{1i} \ E_{2i} \ E_{3i}\end{bmatrix},\\ \begin{bmatrix}\Delta C_{1i} \ \Delta D_{1i}\end{bmatrix} &= H_{2i}F(t)\begin{bmatrix}E_{1i} \ E_{3i}\end{bmatrix},\\ \begin{bmatrix}\Delta C_{2i} \ \Delta D_{2i}\end{bmatrix} &= H_{3i}F(t)\begin{bmatrix}E_{1i} \ E_{2i}\end{bmatrix},\end{aligned}$$

where H_{1i}, H_{2i}, H_{3i}, E_{1i}, E_{2i}, and E_{3i} are known real constant matrices of appropriate dimensions, and $F(t)$ is an unknown matrix function with Lebesgue-measurable elements and satisfies $F(t)^T F(t) \le I$, in which I is the identity matrix of appropriate dimension.

In this chapter, we consider a nonlinear NCS of which the plant is described by the T-S model (11.1). The setup of the overall control system is depicted in Figure 11.1, where $\tau(t) \ge 0$ is the random time-delay from sensor to controller and $\rho(t) \ge 0$ is the random time-delay from controller to actuator. These delays are assumed to be upper bounded. Furthermore, we apply the same Markov processes introduced in Chapter 2 to model the random time-delays in this chapter.

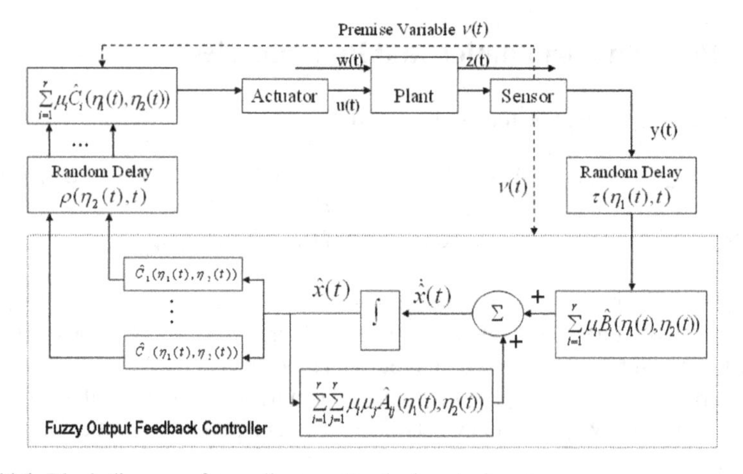

Fig. 11.1 Block diagram of a nonlinear networked control system

In the system setup, the premise vector $\nu(t)$ is connected to the controller and actuators via point-to-point architecture, which is immune to network-induced delays.

Therefore, following the modeling procedure presented in Chapter 2, for the nonlinear plant represented by (11.1), the fuzzy dynamic output feedback controller at time t is inferred as follows:

$$\begin{aligned}\dot{\hat{x}}(t) &= \sum_{i=1}^{r}\sum_{j=1}^{r}\mu_i(v(t))\mu_j(v(t))\Big[\hat{A}_{ij}(\eta_1(t),\eta_2(t))\hat{x}(t)\\&\quad+\hat{B}_i(\eta_1(t),\eta_2(t))y(t-\tau(\eta_1(t),t))\Big],\\u(t) &= \sum_{i=1}^{r}\mu_i(v(t))\hat{C}_i(\eta_1(t),\eta_2(t))\hat{x}(t),\end{aligned} \tag{11.2}$$

where $\hat{A}_{ij}(\eta_1(t),\eta_2(t))$, $\hat{B}_i(\eta_1(t),\eta_2(t))$, $\hat{C}_i(\eta_1(t),\eta_2(t))$ in each plant rule are parameters of the controller which are to be designed. The plant model is rewritten as:

$$\begin{aligned}\dot{x}(t) &= \sum_{i=1}^{r}\mu_i(v(t))[(A_i+\Delta A_i)x(t)+(B_{1i}+\Delta B_{1i})w(t)\\&\quad+(B_{2i}+\Delta B_{2i})u(t-\rho(\eta_2(t),t))],\\z(t) &= \sum_{i=1}^{r}\mu_i(v(t))[(C_{1i}+\Delta C_{1i})x(t)+(D_{1i}+\Delta D_{1i})u(t-\rho(\eta_2(t),t))],\\y(t) &= \sum_{i=1}^{r}\mu_i(v(t))[(C_{2i}+\Delta C_{2i})x(t)+(D_{2i}+\Delta D_{2i})w(t)].\end{aligned} \tag{11.3}$$

Substituting (11.2) into (11.3) yields

$$\begin{aligned}\dot{\tilde{x}}(t) &= \sum_{i=1}^{r}\sum_{j=1}^{r}\mu_i(v(t))\mu_j(v(t))\Big[\tilde{A}_{ij}(\eta_1(t),\eta_2(t))\tilde{x}(t)\\&\quad+\tilde{B}_{ij}(\eta_1(t),\eta_2(t))\tilde{x}(t-\tau(\eta_1(t),t))+\tilde{C}_{ij}(\eta_1(t),\eta_2(t))\tilde{x}(t-\rho(\eta_2(t),t))\\&\quad+\tilde{D}_{ij}(\eta_1(t),\eta_2(t))\omega(t)\Big],\end{aligned} \tag{11.4}$$

where

$$\begin{aligned}\tilde{x}(t) &= \begin{bmatrix} x(t)\\ \hat{x}(t)\end{bmatrix},\quad \omega(t)=\begin{bmatrix} w(t)\\ w(t-\tau(\eta_1(t),t))\end{bmatrix},\\\tilde{A}_{ij}(\eta_1(t),\eta_2(t)) &= \begin{bmatrix} A_i+\Delta A_i & 0\\ 0 & \hat{A}_{ij}(\eta_1(t),\eta_2(t))\end{bmatrix},\\\tilde{B}_{ij}(\eta_1(t),\eta_2(t)) &= \begin{bmatrix} 0 & 0\\ \hat{B}_i(\eta_1(t),\eta_2(t))(C_{2j}+\Delta C_{2j}) & 0\end{bmatrix},\\\tilde{C}_{ij}(\eta_1(t),\eta_2(t)) &= \begin{bmatrix} 0 & (B_{2i}+\Delta B_{2i})\hat{C}_j(\eta_1(t),\eta_2(t))\\ 0 & 0\end{bmatrix},\\\tilde{D}_{ij}(\eta_1(t),\eta_2(t)) &= \begin{bmatrix} B_{1i}+\Delta B_{1i} & 0\\ 0 & \hat{B}_i(\eta_1(t),\eta_2(t))(D_{2j}+\Delta D_{2j})\end{bmatrix}.\end{aligned}$$

The aim of this chapter is to design a dynamic output feedback controller of the form (11.2) such that the following inequality holds:

For (11.4) with its zero state response ($x(\phi)=0$, $\omega(\phi)=0$, $-\chi\le\phi\le 0$),

$$\mathbf{E}\Big[\int_0^{T_f} z^T(t)z(t)dt\Big] \leq \gamma^2 \mathbf{E}\Big[\int_0^{T_f} \sup_{-\chi\leq\phi\leq 0} \omega^T(t+\phi)\omega(t+\phi)dt\Big], \tag{11.5}$$

for any nonzero $\omega(t) \in \mathscr{L}_2[0,\ T_f]$ and $T_f \geq 0$, provided $x = \{x(\xi) : t-2\chi \leq \xi \leq t\} \in L^2_{\mathscr{F}_t}([-2\chi,0];\mathbb{R}^n)$ satisfying:

$$\begin{aligned}&\mathbf{E}\Big[\min_{\eta_1(t)\in\mathscr{S},\eta_2(t)\in\mathscr{W}} V(x(\xi),\eta_1(\xi),\eta_2(\xi),\xi)\Big]\\ &< \delta\mathbf{E}\Big[\max_{\eta_1(t)\in\mathscr{S},\eta_2(t)\in\mathscr{W}} V(x(t),\eta_1(t),\eta_2(t),t)\Big],\end{aligned} \tag{11.6}$$

for all $t-2\chi \leq \xi \leq t$. Here $\mathbf{E}[\cdot]$ stands for the mathematical expectation.

Then the system (11.4) is to be stochastically stabilizable with a disturbance attenuation level γ.

Remark 11.1. From (11.5), it is easy to find that once there is no time-delay in the system, i.e., $\phi = 0$, (11.5) reduces to $\mathbf{E}\Big[\int_0^{T_f} z^T(t)z(t)dt\Big] \leq \gamma^2\mathbf{E}\Big[\int_0^{T_f} \omega^T(t)\omega(t)dt\Big]$, which is $\mathscr{H}_\infty$ control problem.

In this chapter, we assume $u(t) = 0$ before the first control signal reaches the plant. From here, $\mu_i(\nu(t))$ and $\mu_j(\nu(t))$ are denoted as μ_i and μ_j respectively for the convenience of notations. In the symmetric block matrices, we use (*) as an ellipsis for terms that are induced by symmetry. $\hat{A}_{ij}(\eta_1(t),\eta_2(t))$ is denoted as $\hat{A}_{ij}(\iota,\kappa)$ if $\eta_1(t) = \iota$ and $\eta_2(t) = \kappa$.

11.2 Main Result

The following theorem provides sufficient conditions for the existence of a mode-dependent dynamic output feedback controller for the system (11.4) that guarantees disturbance attenuation level γ.

Theorem 11.1. *Consider the system (11.4) satisfying Assumption 11.1. Given positive scalars $\hbar(\iota,\kappa)$, $\varepsilon_{1ij_{\iota\kappa}}$, $\varepsilon_{2ij_{\iota\kappa}}$, $\varepsilon_{3ij_{\iota\kappa}}$, $\varepsilon_{4ij_{\iota\kappa}}$, $\varepsilon_{5ij_{\iota\kappa}}$, $\varepsilon_{6ij_{\iota\kappa}}$, $\varepsilon_{7ij_{\iota\kappa}}$, $\varepsilon_{8ij_{\iota\kappa}}$, and $\varepsilon_{9ij_{\iota\kappa}}$, if there exist symmetric matrices $X(\iota,\kappa)$, $Y(\iota,\kappa)$, matrices $L_i(\iota,\kappa)$, $F_i(\iota,\kappa)$, and positive scalars $\beta_{1_{\iota\kappa}}$, $\beta_{2_{\iota\kappa}}$, such that the following inequalities hold for all $\iota \in \mathscr{S}$ and $\kappa \in \mathscr{W}$:*

$$\begin{bmatrix} Y(\iota,\kappa) & I \\ I & X(\iota,\kappa) \end{bmatrix} > 0, \tag{11.7}$$

$$\Omega_{ii}(\iota,\kappa) < 0, \textit{for } i \in \mathscr{I}_R \tag{11.8}$$

$$\Omega_{ij}(\iota,\kappa) + \Omega_{ji}(\iota,\kappa) < 0, \textit{for } i < j < r \tag{11.9}$$

$$\Upsilon_{ij}(\iota,\kappa) < 0, \textit{for } \{i,j\} \in \mathscr{I}_R \times \mathscr{I}_R \tag{11.10}$$

$$\begin{bmatrix} R_{4_{\iota\kappa}} & (*)^T \\ \Lambda_\iota^T & \mathscr{Q}_{1_{\iota\kappa}} \end{bmatrix} > 0, \tag{11.11}$$

$$\begin{bmatrix} R_{5_{\iota\kappa}} & (*)^T \\ \Pi_\kappa^T & \mathscr{Q}_{2_{\iota\kappa}} \end{bmatrix} > 0, \tag{11.12}$$

$$\begin{bmatrix} -R_{1_{\iota\kappa}} & (*)^T & (*)^T & (*)^T \\ 0 & -I & (*)^T & (*)^T \\ 0 & -Y(\iota,\kappa) & -R_{2_{\iota\kappa}} & (*)^T \\ 0 & 0 & 0 & -R_{3_{\iota\kappa}} \end{bmatrix} < 0, \tag{11.13}$$

$$\begin{bmatrix} -\beta_{2_{\iota\kappa}} Y(\iota,\kappa) & (*)^T & (*)^T & (*)^T & (*)^T & (*)^T \\ -\beta_{2_{\iota\kappa}} I & -\beta_{2_{\iota\kappa}} X(\iota,\kappa) & (*)^T & (*)^T & (*)^T & (*)^T \\ L_j^T(\iota,\kappa) B_i^T & L_j^T(\iota,\kappa) B_i^T X(\iota,\kappa) & -Y(\iota,\kappa) & (*)^T & (*)^T & (*)^T \\ 0 & 0 & -I & -X(\iota,\kappa) & (*)^T & (*)^T \\ \varepsilon_{6ij_{\iota\kappa}} H_{1i}^T & \varepsilon_{6ij_{\iota\kappa}} H_{1i}^T X(\iota,\kappa) & 0 & 0 & -\varepsilon_{6ij_{\iota\kappa}} I & (*)^T \\ 0 & 0 & E_{2i} L_j(\iota,\kappa) & 0 & 0 & -\varepsilon_{6ij_{\iota\kappa}} I \end{bmatrix} < 0,$$

$$\textit{for } \{i,j\} \in \mathscr{I}_R \times \mathscr{I}_R \tag{11.14}$$

$$\begin{bmatrix} -\beta_{2_{\iota\kappa}} Y(\iota,\kappa) & (*)^T & (*)^T & (*)^T & (*)^T & (*)^T \\ -\beta_{2_{\iota\kappa}} I & -\beta_{2_{\iota\kappa}} X(\iota,\kappa) & (*)^T & (*)^T & (*)^T & (*)^T \\ 0 & Y(\iota,\kappa) C_{2j}^T F_i^T(\iota,\kappa) & -Y(\iota,\kappa) & (*)^T & (*)^T & (*)^T \\ 0 & 0 & -I & -X(\iota,\kappa) & (*)^T & (*)^T \\ 0 & \varepsilon_{7ij_{\iota\kappa}} H_{2j}^T F_i^T(\iota,\kappa) & 0 & 0 & -\varepsilon_{7ij_{\iota\kappa}} I & (*)^T \\ 0 & 0 & E_{1j} Y(\iota,\kappa) & 0 & 0 & -\varepsilon_{7ij_{\iota\kappa}} I \end{bmatrix} < 0,$$

$$\textit{for } \{i,j\} \in \mathscr{I}_R \times \mathscr{I}_R \tag{11.15}$$

$$\begin{bmatrix} -Y(\iota,\kappa) & (*)^T & (*)^T & (*)^T & (*)^T & (*)^T & (*)^T & (*)^T \\ -I & -X(\iota,\kappa) & (*)^T & (*)^T & (*)^T & (*)^T & (*)^T & (*)^T \\ B_{1i}^T & B_{1i}^T X(\iota,\kappa) & -I & (*)^T & (*)^T & (*)^T & (*)^T & (*)^T \\ 0 & D_{2i}^T F_j^T(\iota,\kappa) & 0 & -I & (*)^T & (*)^T & (*)^T & (*)^T \\ \varepsilon_{8ij_{\iota\kappa}} H_{1i}^T & \varepsilon_{8ij_{\iota\kappa}} H_{1i}^T X(\iota,\kappa) & 0 & 0 & -\varepsilon_{8ij_{\iota\kappa}} I & (*)^T & (*)^T & (*)^T \\ 0 & 0 & E_{2i} & 0 & 0 & -\varepsilon_{8ij_{\iota\kappa}} I & (*)^T & (*)^T \\ 0 & \varepsilon_{9ij_{\iota\kappa}} H_{3i}^T F_j^T(\iota,\kappa) & 0 & 0 & 0 & 0 & -\varepsilon_{9ij_{\iota\kappa}} I & (*)^T \\ 0 & 0 & 0 & E_{2i} & 0 & 0 & 0 & -\varepsilon_{9ij_{\iota\kappa}} I \end{bmatrix} < 0,$$

$$\textit{for } \{i,j\} \in \mathscr{I}_R \times \mathscr{I}_R \tag{11.16}$$

where

$$\Upsilon_{ij}(\iota,\kappa)=\begin{bmatrix} -\beta_{1_{\iota\kappa}}Y(\iota,\kappa)+2R_{1_{\iota\kappa}} & (*)^T & (*)^T & (*)^T & (*)^T & (*)^T & (*)^T & (*)^T \\ -\beta_{1_{\iota\kappa}}I & -\beta_{1_{\iota\kappa}}X(\iota,\kappa) & (*)^T & (*)^T & (*)^T & (*)^T & (*)^T & (*)^T \\ Y(\iota,\kappa)A_i^T & \begin{pmatrix} -A_i-L_j^T(\iota,\kappa)B_{2i}^TX(\iota,\kappa) \\ -Y(\iota,\kappa)C_{2j}^TF_i^T(\iota,\kappa) \\ -(\lambda_{\iota\iota}+\pi_{\kappa\kappa})I \end{pmatrix} & -Y(\iota,\kappa)+2R_{2_{\iota\kappa}} & (*)^T & (*)^T & (*)^T & (*)^T & (*)^T \\ A_i^T & A_i^TX(\iota,\kappa) & -I & -X(\iota,\kappa)+2R_{3_{\iota\kappa}} & (*)^T & (*)^T & (*)^T & (*)^T \\ 0 & R_{4_{\iota\kappa}} & 0 & 0 & -I & (*)^T & (*)^T & (*)^T \\ 0 & R_{5_{\iota\kappa}} & 0 & 0 & 0 & -I & (*)^T & (*)^T \\ \varepsilon_{5ij_{\iota\kappa}}H_{1i}^T & \varepsilon_{5ij_{\iota\kappa}}H_{1i}^TX(\iota,\kappa) & 0 & 0 & 0 & 0 & -\varepsilon_{5ij_{\iota\kappa}}I & (*)^T \\ 0 & 0 & E_{1i}Y(\iota,\kappa) & E_{1i} & 0 & 0 & 0 & -\varepsilon_{5ij_{\iota\kappa}}I \end{bmatrix}$$

$$\Omega_{ij}(\iota,\kappa)=\begin{bmatrix} \Xi_{1ij}(\iota,\kappa) & (*)^T & (*)^T & (*)^T & (*)^T & (*)^T & (*)^T & (*)^T & (*)^T & (*)^T & (*)^T & (*)^T & (*)^T & (*)^T & (*)^T \\ (\beta_{1_{\iota\kappa}}+6\beta_{2_{\iota\kappa}})\hbar_{\iota\kappa}I & \Xi_{2ij}(\iota,\kappa) & (*)^T & (*)^T & (*)^T & (*)^T & (*)^T & (*)^T & (*)^T & (*)^T & (*)^T & (*)^T & (*)^T & (*)^T & (*)^T \\ B_{1i}^T & B_{1i}^TX(\iota,\kappa) & -\gamma_{d_f}I & (*)^T & (*)^T & (*)^T & (*)^T & (*)^T & (*)^T & (*)^T & (*)^T & (*)^T & (*)^T & (*)^T & (*)^T \\ 0 & D_{2i}^TF_j^T(\iota,\kappa) & 0 & -\gamma_{d_f}I & (*)^T & (*)^T & (*)^T & (*)^T & (*)^T & (*)^T & (*)^T & (*)^T & (*)^T & (*)^T & (*)^T \\ C_{1i}Y(\iota,\kappa)+D_{1i}L_j(\iota,\kappa) & C_{1i} & 0 & 0 & -I & (*)^T & (*)^T & (*)^T & (*)^T & (*)^T & (*)^T & (*)^T & (*)^T & (*)^T & (*)^T \\ E_{1i}Y(\iota,\kappa)+E_{3i}L_j(\iota,\kappa) & 0 & E_{2i} & 0 & 0 & -\varepsilon_{1ij_{\iota\kappa}}I & (*)^T & (*)^T & (*)^T & (*)^T & (*)^T & (*)^T & (*)^T & (*)^T & (*)^T \\ \varepsilon_{1ij_{\iota\kappa}}H_{1i}^T & 0 & 0 & 0 & 0 & 0 & -\varepsilon_{1ij_{\iota\kappa}}I & (*)^T & (*)^T & (*)^T & (*)^T & (*)^T & (*)^T & (*)^T & (*)^T \\ E_{1i}Y(\iota,\kappa) & E_{1i} & E_{2i} & 0 & 0 & 0 & 0 & -\varepsilon_{2ij_{\iota\kappa}}I & (*)^T & (*)^T & (*)^T & (*)^T & (*)^T & (*)^T & (*)^T \\ 0 & \varepsilon_{2ij_{\iota\kappa}}H_{1i}^TX(\iota,\kappa) & 0 & 0 & 0 & 0 & 0 & 0 & -\varepsilon_{2ij_{\iota\kappa}}I & (*)^T & (*)^T & (*)^T & (*)^T & (*)^T & (*)^T \\ 0 & E_{1i} & 0 & E_{2i} & 0 & 0 & 0 & 0 & 0 & -\varepsilon_{3ij_{\iota\kappa}}I & (*)^T & (*)^T & (*)^T & (*)^T & (*)^T \\ 0 & \varepsilon_{3ij_{\iota\kappa}}H_{3i}^TF_j^T(\iota,\kappa) & 0 & 0 & 0 & 0 & 0 & 0 & 0 & 0 & -\varepsilon_{3ij_{\iota\kappa}}I & (*)^T & (*)^T & (*)^T & (*)^T \\ E_{1i}Y(\iota,\kappa)+E_{3i}L_j(\iota,\kappa) & E_{1i} & 0 & 0 & 0 & 0 & 0 & 0 & 0 & 0 & 0 & -\varepsilon_{4ij_{\iota\kappa}}I & (*)^T & (*)^T & (*)^T \\ 0 & 0 & 0 & 0 & \varepsilon_{4ij_{\iota\kappa}}H_{2i}^T & 0 & 0 & 0 & 0 & 0 & 0 & 0 & -\varepsilon_{4ij_{\iota\kappa}}I & (*)^T & (*)^T \\ S^T(\iota,\kappa) & 0 & 0 & 0 & 0 & 0 & 0 & 0 & 0 & 0 & 0 & 0 & 0 & -\mathscr{Q}_{1_{\iota\kappa}} & (*)^T \\ Z^T(\iota,\kappa) & 0 & 0 & 0 & 0 & 0 & 0 & 0 & 0 & 0 & 0 & 0 & 0 & 0 & -\mathscr{Q}_{2_{\iota\kappa}} \end{bmatrix}$$

with

$$\Xi_{1ij}(\iota,\kappa) = A_iY(\iota,\kappa)+Y(\iota,\kappa)A_i^T + B_{2i}L_j(\iota,\kappa)+L_j^T(\iota,\kappa)B_{2i}^T$$
$$+(\beta_{1_{\iota\kappa}}+6\beta_{2_{\iota\kappa}})\hbar_{\iota\kappa}Y(\iota,\kappa)+(\lambda_{\iota\iota}+\pi_{\kappa\kappa})Y(\iota,\kappa)$$
$$\Xi_{2ij}(\iota,\kappa) = X(\iota,\kappa)A_i+A_i^TX(\iota,\kappa)+F_i(\iota,\kappa)C_{2j}+C_{2j}^TF_i^T(\iota,\kappa)$$
$$+(\beta_{1_{\iota\kappa}}+6\beta_{2_{\iota\kappa}})\hbar_{\iota\kappa}X(\iota,\kappa)+\sum_{\wp=1}^{s}\lambda_{\iota\wp}X(\wp,\kappa)+\sum_{\ell=1}^{w}\pi_{\kappa\ell}X(\iota,\ell)$$

and

$$S(\iota,\kappa) = [\sqrt{\lambda_{\iota 1}}Y(\iota,\kappa)\cdots\sqrt{\lambda_{\iota(\iota-1)}}Y(\iota,\kappa)\ \sqrt{\lambda_{\iota(\iota+1)}}Y(\iota,\kappa)\cdots\sqrt{\lambda_{\iota s}}Y(\iota,\kappa)],$$
$$Z(\iota,\kappa) = [\sqrt{\pi_{\kappa 1}}Y(\iota,\kappa)\cdots\sqrt{\pi_{\kappa(\kappa-1)}}Y(\iota,\kappa)\ \sqrt{\pi_{\kappa(\kappa+1)}}Y(\iota,\kappa)\cdots\sqrt{\pi_{\kappa w}}Y(\iota,\kappa)],$$
$$\Lambda_\iota = [\sqrt{\lambda_{\iota 1}}I\cdots\sqrt{\lambda_{\iota(\iota-1)}}I\ \sqrt{\lambda_{\iota(\iota+1)}}I\cdots\sqrt{\lambda_{\iota s}}I],$$
$$\Pi_\kappa = [\sqrt{\pi_{\kappa 1}}I\cdots\sqrt{\pi_{\kappa(\kappa-1)}}I\ \sqrt{\pi_{\kappa(\kappa+1)}}I\cdots\sqrt{\pi_{\kappa w}}I],$$
$$\mathscr{Q}_{1_{\iota\kappa}} = diag\{Y(1,\kappa),\cdots,Y(\iota-1,\kappa),Y(\iota+1,\kappa),\cdots,Y(s,\kappa)\},$$
$$\mathscr{Q}_{2_{\iota\kappa}} = diag\{Y(\iota,1),\cdots,Y(\iota,\kappa-1),Y(\iota,\kappa+1),\cdots,Y(\iota,w)\},$$

then the system (11.4) is said to achieve stochastic stability via controller (11.2) for all delays $\tau(\iota,t)$ and $\rho(\kappa,t)$ satisfying

$$0\le\tau(\iota,t)+\rho(\kappa,t)\le\hbar(\iota,\kappa).$$

Furthermore, the mode dependant controller is of the form (11.2) with

$$\hat{A}_{ij}(\iota,\kappa) = [Y^{-1}(\iota,\kappa)-X(\iota,\kappa)]^{-1}[-A_i^T-X(\iota,\kappa)A_iY(\iota,\kappa)-F_i(\iota,\kappa)C_jY(\iota,\kappa)$$
$$-X(\iota,\kappa)B_iL_j(\iota,\kappa)-\sum_{\wp=1}^{s}\lambda_{\iota\wp}Y^{-1}(\wp,\kappa)Y(\iota,\kappa)$$
$$-\sum_{\ell=1}^{w}\pi_{\kappa\ell}Y^{-1}(\iota,\ell)Y(\iota,\kappa)]Y^{-1}(\iota,\kappa), \tag{11.17}$$
$$\hat{B}_i(\iota,\kappa) = [Y^{-1}(\iota,\kappa)-X(\iota,\kappa)]^{-1}F_i(\iota,\kappa), \tag{11.18}$$
$$\hat{C}_i(\iota,\kappa) = L_i(\iota,\kappa)Y^{-1}(\iota,\kappa). \tag{11.19}$$

Proof. The results can be obtained straightforward from the proof process in Chapter 9 and Chapter 5. □

The iterative algorithm presented in Chapter 9 is applied here so solve Theorem 11.1 which is a BMI problem.

11.3 Numerical Example

To illustrate the validation of the results obtained previously, we consider the following problem of balancing an inverted pendulum on a cart. The equations of motion of the pendulum are described as follows:

$$\dot{x_1} = x_2,$$
$$\dot{x_2} = \frac{g\sin(x_1) - amlx_2^2\sin(2x_1)/2 - a\cos(x_1)u}{4l/3 - aml\cos^2(x_1)} + w, \tag{11.20}$$

where x_1 denotes the angle of the pendulum from the vertical position, and x_2 is the angular velocity. $g = 9.8m/s^2$ is the gravity constant, m is the mass of the pendulum, $a = 1/(m+M)$, M is the mass of the cart, $2l$ is the length of the pendulum, and u is the force applied to the cart. In the simulation, the pendulum parameters are chosen as $m = 2kg$, $M = 8kg$, and $2l = 1.0m$.

We approximate the system (11.20) by the following T-S fuzzy model:

$$Rule\ 1: If\ x_1(t)\ is\ M_1,\ then$$
$$\dot{x}(t) = (A_1 + \Delta A_1)x(t) + B_1 w(t) + (B_{2_1} + \Delta B_{2_1})u(t)$$
$$z(t) = C_1 x(t) + D_{12} u(t)$$
$$y(t) = C_2 x(t)$$

$$Rule\ 2: If\ x_1(t)\ is\ M_2,\ then$$
$$\dot{x}(t) = (A_2 + \Delta A_2)x(t) + B_1 w(t) + (B_{2_2} + \Delta B_{2_2})u(t)$$
$$z(t) = C_1 x(t) + D_{12} u(t)$$
$$y(t) = C_2 x(t)$$

where

$$A_1 = \begin{bmatrix} 0 & 1 \\ \frac{g}{4l/3 - aml} & 0 \end{bmatrix}, B_{2_1} = \begin{bmatrix} 0 \\ -\frac{a}{4l/3 - aml} \end{bmatrix},$$
$$A_2 = \begin{bmatrix} 0 & 1 \\ \frac{2g}{\pi(4l/3 - aml\beta^2)} & 0 \end{bmatrix}, B_{2_2} = \begin{bmatrix} 0 \\ -\frac{a\beta}{4l/3 - aml\beta^2} \end{bmatrix},$$
$$B_1 = \begin{bmatrix} 0 \\ 1 \end{bmatrix}, C_1 = \begin{bmatrix} 1 & 0.3 \end{bmatrix}, D_{12} = 0.01, C_2 = \begin{bmatrix} 9 & 0.1 \end{bmatrix},$$
$$H_{1_1} = H_{1_2} = \begin{bmatrix} 0.3 & 0 \\ 0 & 0.3 \end{bmatrix}, E_{1_1} = E_{1_2} = \begin{bmatrix} 0.5 & 0 \\ 0 & 0.5 \end{bmatrix}, E_{2_1} = E_{2_2} = \begin{bmatrix} 0 \\ 0.2 \end{bmatrix},$$

and $\beta = \cos(88^\circ)$. The disturbance attenuation level γ is set to be equal to 1 in this example and $\varepsilon_1 = \varepsilon_2 = 1$. The membership functions for Rule 1 and Rule 2 are shown in Figure 11.2.

In our simulation, we assume the sampling period is 0.005 for both sensor and actuation channels, that is, $h^a = h^s = 0.005$, and $n^s = n^a = 0$ which means no data packet dropout happens in the communication channel. Delay free attenuation constant γ_{d_f} is set to be 1, while constants $\varepsilon_{1ij_{\iota\kappa}}$, $\varepsilon_{2ij_{\iota\kappa}}$, $\varepsilon_{3ij_{\iota\kappa}}$, $\varepsilon_{4ij_{\iota\kappa}}$, $\varepsilon_{5ij_{\iota\kappa}}$, $\varepsilon_{6ij_{\iota\kappa}}$, $\varepsilon_{7ij_{\iota\kappa}}$, $\varepsilon_{8ij_{\iota\kappa}}$, and $\varepsilon_{9ij_{\iota\kappa}}$ are set be equivalent to 1.

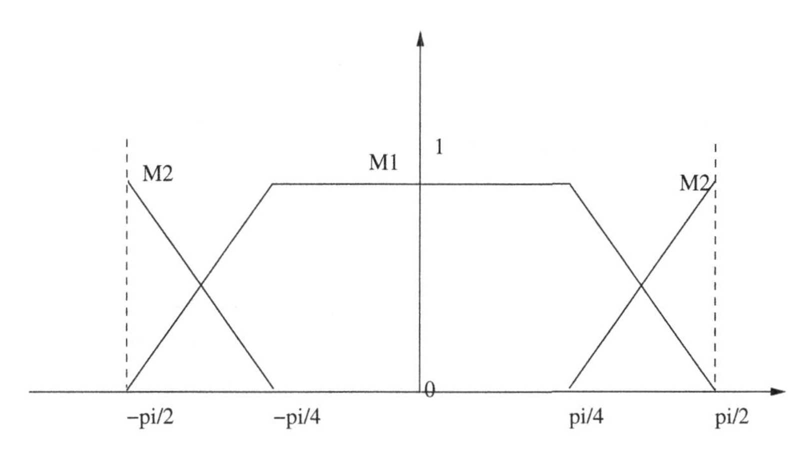

Fig. 11.2 Membership function

In the following simulation, we assume $F(t) = \sin t$ and it can be seen that $\|F(t)\| \leq 1$.

The random time-delays exist in $\mathscr{S} = \{1,2\}$ and $\mathscr{W} = \{1,2\}$, and their transition rate matrices are given by:

$$\Lambda = \begin{bmatrix} -3 & 3 \\ 2 & -2 \end{bmatrix}, \Pi = \begin{bmatrix} -1 & 1 \\ 2 & -2 \end{bmatrix}.$$

Furthermore, we assume that the sensor-to-controller communication delays for two Markovian modes are $|\tau_1^s| < 0.01$, $|\tau_2^s| < 0.008$, while the controller-to-actuator delays are $|\tau_1^a| < 0.007$, and $|\tau_2^a| < 0.012$, and therefore by (2.6) and (2.7) we can have $\hbar_{11} = 0.027$, $\hbar_{12} = 0.032$, $\hbar_{21} = 0.025$, and $\hbar_{22} = 0.03$. By applying Theorem 11.1 and the iterative algorithm, we get the following controller gains by the calculation of (11.17)-(11.19):

$$\hat{A}_{11}(1,1) = \begin{bmatrix} -20.589 & -56.924 \\ 172.94 & -9.0175 \end{bmatrix}, \hat{A}_{12}(1,1) = \begin{bmatrix} -21.434 & -55.46 \\ 175.543 & -9.554 \end{bmatrix},$$

$$\hat{A}_{21}(1,1) = \begin{bmatrix} -20.964 & -53.3222 \\ 170.22 & -10.21 \end{bmatrix}, \hat{A}_{22}(1,1) = \begin{bmatrix} -21.2219 & -57.14 \\ 169.58 & -9.15 \end{bmatrix},$$

$$\hat{B}_1(1,1) = \begin{bmatrix} 1.9634 \\ -3.0617 \end{bmatrix}, \hat{B}_2(1,1) = \begin{bmatrix} 4.4349 \\ -9.6547 \end{bmatrix},$$

$$\hat{C}_1(1,1) = \begin{bmatrix} 3.4116 & -9.9555 \end{bmatrix}, \hat{C}_2(1,1) = \begin{bmatrix} 2.9754 & -10.5743 \end{bmatrix},$$

$$\hat{A}_{11}(1,2) = \begin{bmatrix} -34.519 & -57.438 \\ 226.33 & -5.0388 \end{bmatrix}, \hat{A}_{12}(1,2) = \begin{bmatrix} -35.002 & -59.545 \\ 254.678 & -5.22 \end{bmatrix},$$
$$\hat{A}_{21}(1,2) = \begin{bmatrix} -39.543 & -56.412 \\ 244.4906 & -6.7648 \end{bmatrix}, \hat{A}_{22}(1,2) = \begin{bmatrix} -38.719 & -58.087 \\ 245.234 & -7.031 \end{bmatrix},$$
$$\hat{B}_1(1,2) = \begin{bmatrix} 2.3327 \\ -4.4194 \end{bmatrix}, \hat{B}_2(1,2) = \begin{bmatrix} 5.6578 \\ -6.535 \end{bmatrix},$$
$$\hat{C}_1(1,2) = \begin{bmatrix} 3.3543 & -9.844 \end{bmatrix}, \hat{C}_2(1,2) = \begin{bmatrix} 1.9994 & -5.541 \end{bmatrix},$$
$$\hat{A}_{11}(2,1) = \begin{bmatrix} -14.197 & -56.603 \\ 148.01 & -10.9 \end{bmatrix}, \hat{A}_{12}(2,1) = \begin{bmatrix} -15.4276 & -55.434 \\ 143.998 & -10.843 \end{bmatrix},$$
$$\hat{A}_{21}(2,1) = \begin{bmatrix} -17.095 & -59.483 \\ 145.321 & -11.5743 \end{bmatrix}, \hat{A}_{22}(2,1) = \begin{bmatrix} -16.6546 & -57.9798 \\ 148.01 & -10.439 \end{bmatrix},$$
$$\hat{B}_1(2,1) = \begin{bmatrix} 1.7708 \\ -2.4243 \end{bmatrix}, \hat{B}_2(2,1) = \begin{bmatrix} 2.584 \\ -6.49 \end{bmatrix},$$
$$\hat{C}_1(2,1) = \begin{bmatrix} 3.3808 & -10.01 \end{bmatrix}, \hat{C}_2(2,1) = \begin{bmatrix} 3.3789 & -12.6654 \end{bmatrix},$$
$$\hat{A}_{11}(2,2) = \begin{bmatrix} -23.167 & -56.865 \\ 182.63 & -8.2498 \end{bmatrix}, \hat{A}_{12}(2,2) = \begin{bmatrix} -27.4833 & -54.238 \\ 187.493 & -10.4738 \end{bmatrix},$$
$$\hat{A}_{21}(2,2) = \begin{bmatrix} -25.096 & -56.0943 \\ 181.48 & -9.924 \end{bmatrix}, \hat{A}_{22}(2,2) = \begin{bmatrix} -24.8540 & -55.5496 \\ 189.9403 & -8.933 \end{bmatrix},$$
$$\hat{B}_1(2,2) = \begin{bmatrix} 2.0131 \\ -3.2839 \end{bmatrix}, \hat{B}_2(2,2) = \begin{bmatrix} 2.438 \\ -3.3444 \end{bmatrix},$$
$$\hat{C}_1(2,2) = \begin{bmatrix} 3.3512 & -9.9144 \end{bmatrix}, \hat{C}_2(2,2) = \begin{bmatrix} 5.0433 & -5.9403 \end{bmatrix}.$$

The ratio of the regulated output energy to the disturbance input noise is depicted in Figure 11.3. In our simulation, we use a uniform distributed random disturbance input signal $w(t)$ with maximum value 2. It can be seen that the ratio tends to a constant value of about 0.05, which means the attenuation level equals to $\sqrt{0.05} \approx 0.22$, less than the prescribed level $\gamma = \sqrt{\gamma_{d_f} + \max(\hbar_{\iota\kappa})} = \sqrt{1+0.032} \approx 1.016$.

In conclusion, the designed controller meets the performance requirements.

11.4 Conclusion

In this chapter, a technique of designing a delay-dependant dynamic output feedback controller with robust disturbance attenuation and stability for an uncertain nonlinear NCS has been proposed. The delays in the communication network are regarded as input delays and are dealt with in the scope of disturbance attenuation. The Lyapunov–Razumikhin method has been employed to derive such a controller for this class of systems. Sufficient conditions for the existence of such a controller for this class of NCSs are derived. We finally use a numerical example to demonstrate the effectiveness of this methodology in the last section.

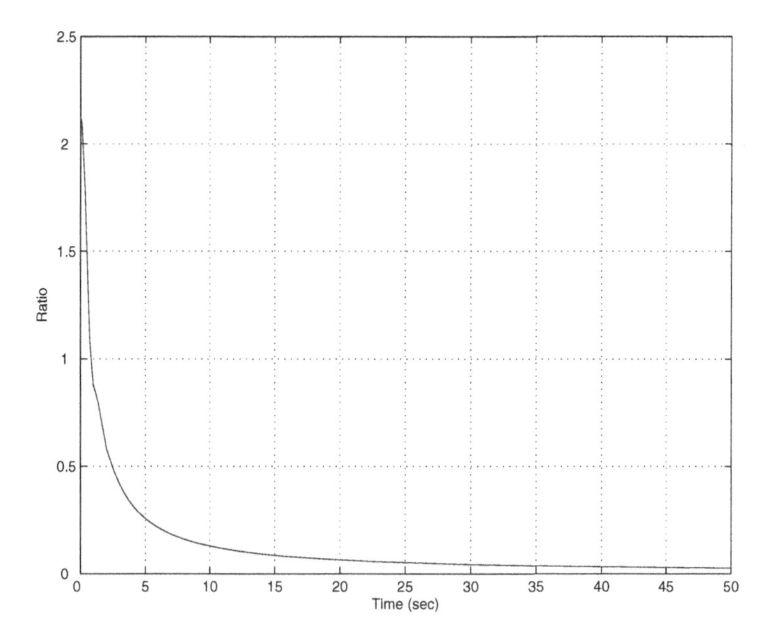

Fig. 11.3 The ratio of the regulated output energy to the disturbance input noise without data dropouts

Chapter 12
Robust Fuzzy Filter Design for Uncertain Nonlinear Networked Control Systems

This chapter investigates the problem of robust fuzzy filter design for a class of nonlinear NCSs, which is a development of the results obtained in Chapter 6. The nonlinear plant is firstly represented by a set of local linear models based on the T-S fuzzy modeling technique presented in Chapter 7. We then proceed to the design of robust fuzzy filters for the system. By applying Lyapunov–Razumikhin method, the existence of a delay-dependent filter is given in terms of the solvability of BMIs. It should be noted that the overall designed fuzzy filter is also a blending of local linear filters according to a given fuzzy rules. The viability of the results is verified by a real world example.

12.1 Problem Formulation and Preliminaries

In this chapter, we assume that $u(t) = 0$ without loss of generality. By using the T-S fuzzy modeling procedure presented in Chapter 8, the plant dynamics of the uncertain nonlinear system is described as follows:

$$\begin{cases} \dot{x}(t) = \sum_{i=1}^{r} \mu_i(v(t))[(A_i + \Delta A_i)x(t) + (B_{1i} + \Delta B_{1i})w(t)] \\ z(t) = \sum_{i=1}^{r} \mu_i(v(t))[(C_{1i} + \Delta C_{1i})x(t)] \\ y(t) = \sum_{i=1}^{r} \mu_i(v(t))[(C_{2i} + \Delta C_{2i})x(t) + (D_{2i} + \Delta D_{2i})w(t)] \end{cases} \tag{12.1}$$

where $x(t) \in \mathbb{R}^n$ is the state vector, $w(t) \in \mathbb{R}^p$ is the exogenous disturbance input and/or measurement noise, $y(t) \in \mathbb{R}^l$ and $z(t) \in \mathbb{R}^s$ denote the measurement and regulated output respectively.

Furthermore, $i \in \mathscr{I}_R = \{1, \cdots, r\}$, r is the number of fuzzy rules; $v_k(t)$ are premise variables, M_{ik} are fuzzy sets, $k = 1, \cdots, p$, p is the number of premise variables

$$v(t) = [v_1(t), v_2(t), \cdots, v_p(t)]^T,$$

D. Huang and S.K. Nguang: Robust Ctrl. for Uncertain Networked Ctrl. Sys., LNCIS 386, pp. 129–136.
springerlink.com

$$\omega_i(v(t)) = \prod_{k=1}^{p} M_{ik}(v_k(t)),\ \omega_i(v(t)) \geq 0,\ \sum_{i=1}^{r} \omega_i(v(t)) > 0,$$

$$\mu_i(v(t)) = \frac{\omega_i(v(t))}{\sum_{i=1}^{r} \omega_i(v(t))},\ \mu_i(v(t)) \geq 0,\ \sum_{i=1}^{r} \mu_i(v(t)) = 1.$$

Here, $M_{ik}(v_k(t))$ denote the grade of membership of $v_k(t)$ in M_{ik}.

Matrices ΔA_{1i}, ΔB_{1i}, ΔC_{1i}, ΔC_{2i}, and ΔD_{2i} characterize the uncertainties in the system and satisfy the following assumption:

Assumption 12.1.

$$\begin{aligned} \left[\Delta A_i\ \Delta B_{1i}\right] &= H_{1i}F(t)\left[E_{1i}\ E_{2i}\right], \\ \Delta C_{1i} &= H_{2i}F(t)E_{1i}, \\ \left[\Delta C_{2i}\ \Delta D_{2i}\right] &= H_{3i}F(t)\left[E_{1i}\ E_{2i}\right], \end{aligned}$$

where H_{1i}, H_{2i}, H_{3i}, E_{1i}, and E_{2i} are known real constant matrices of appropriate dimensions, and $F(t)$ is an unknown matrix function with Lebesgue-measurable elements and satisfies $F(t)^T F(t) \leq I$, in which I is the identity matrix of appropriate dimension.

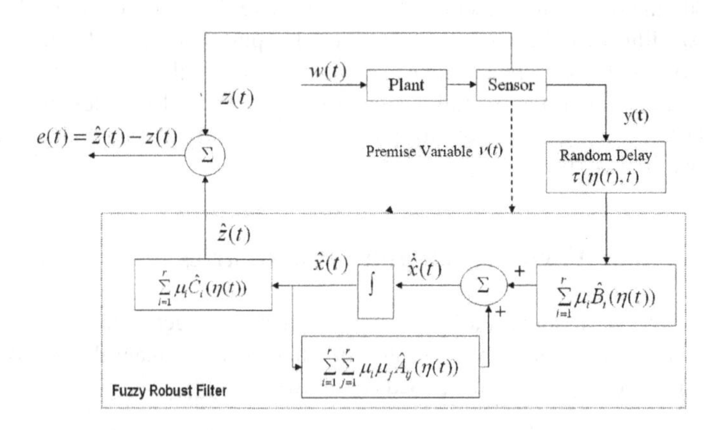

Fig. 12.1 Block diagram of an uncertain nonlinear NCS with a robust filter

In this chapter, we consider a nonlinear NCS with a robust fuzzy filter of which the setup is depicted in Figure 12.1. With the insertion of the communication network, this inevitably introduces complexity to the analysis and design of the system. The main issue is the network-induced effects, i.e., random network delays and data packet dropouts.

Following the same modeling procedure of the aforementioned network-induced effects as detailed in Chapter 2, our aim of this chapter is to design a full order dynamic $\mathscr{H}_\infty$ fuzzy filter of the form:

$$\mathscr{F}:\begin{cases}\dot{\hat{x}}(t)=\sum_{i=1}^{r}\sum_{j=1}^{r}\mu_i(v(t))\mu_j(v(t))[\hat{A}_{ij}(\iota)\hat{x}(t)+\hat{B}_i(\iota)y(t-\tau(t))],\\ \hat{z}(t)=\sum_{i=1}^{r}\mu_i(v(t))\hat{C}_i(\iota)\hat{x}(t),\end{cases}\tag{12.2}$$

Matrices $\hat{A}_{ij}(\iota)$, $\hat{B}_i(\iota)$, $\hat{C}_i(\iota)$ are the filter's parameters.

Therefore, with regard to (12.1) and (12.2), they can be written in the following concise form:

$$\begin{aligned}\dot{\tilde{x}}(t) &= \sum_{i=1}^{r}\sum_{j=1}^{r}\mu_i(v(t))\mu_j(v(t))\Big[\mathscr{A}_{ij}(\eta(t))\tilde{x}(t)\\ &\quad+\mathscr{B}_{ij}(\eta(t))\tilde{x}(t-\tau(\eta(t),t))\\ &\quad+\mathscr{C}_{ij}(\eta(t))\omega(t)\Big],\end{aligned}\tag{12.3}$$

$$e(t)=z(t)-\hat{z}(t)=\sum_{i=1}^{r}\mu_i(v(t))\mathscr{D}_i(\eta(t))\tilde{x}(t),\tag{12.4}$$

where $\tilde{x}(t)=[x^T(t)\ \ \hat{x}^T(t)]^T$, $w(t-\tau(\eta(t),t))=v(t)$, $\omega(t)=[w^T(t)\ \ v^T(t)]^T$, and

$$\begin{aligned}\mathscr{A}_{ij}(\eta(t)) &= \begin{bmatrix}A_i+\Delta A_i & 0\\ 0 & \hat{A}_{ij}(\eta(t))\end{bmatrix},\\ \mathscr{B}_{ij}(\eta(t)) &= \begin{bmatrix}0 & 0\\ \hat{B}_i(\eta(t))(C_{2j}+\Delta C_{2j}) & 0\end{bmatrix},\\ \mathscr{C}_{ij}(\eta(t)) &= \begin{bmatrix}B_{1i}+\Delta B_{1i} & 0\\ 0 & \hat{B}_i(\eta(t))(D_{2j}+\Delta D_{2j})\end{bmatrix},\\ \mathscr{D}_i(\eta(t)) &= \begin{bmatrix}C_{1i}+\Delta C_{1i} & -\hat{C}_i(\eta(t))\end{bmatrix}.\end{aligned}$$

The aim of this chapter is to design a fuzzy filter of the form (12.2) such that the following conditions are met.

Let ζ,α_1,α_2 be all positive numbers and $\delta>1$. Assume that there exists a function $V\in C^{2,1}(\mathbb{R}^n\times\mathscr{S}\times[-\chi,\infty);\mathbb{R}_+)$ such that

$$\begin{aligned}&\alpha_1\|x(t)\|^2\le V(x(t),\eta(t),t)\le\alpha_2\|x(t)\|^2\\ &\text{for all } (x(t),\eta(t),t)\in\mathbb{R}^n\times\mathscr{S}\times\mathscr{W}\times[-\chi,\infty),\end{aligned}\tag{12.5}$$

and also for system (12.1), if its zero state response ($x(\phi)=0,\ \omega(\phi)=0,\ -\chi\le\phi\le0$) with filter (12.2) satisfies,

$$\begin{aligned}&\mathbf{E}\Big[\int_0^{T_f}\Big(z(t)-\hat{z}(t)\Big)^T\Big(z(t)-\hat{z}(t)\Big)dt\Big]\\ &\le\gamma^2\mathbf{E}\Big[\int_0^{T_f}\sup_{-\chi\le\phi\le0}\omega^T(t+\phi)\omega(t+\phi)dt\Big]\end{aligned}\tag{12.6}$$

for any nonzero $\omega(t)\in\mathscr{L}_2[0,\ T_f]$ and $T_f\ge0$, provided $x=\{x(\xi):t-2\chi\le\xi\le t\}\in L^2_{\mathscr{F}_t}([-2\chi,0];\mathbb{R}^n)$ satisfying:

$$\mathbf{E}\Big[\min_{\eta(t)\in\mathscr{S}\in\mathscr{W}} V(x(\xi),\eta(\xi),\xi)\Big] < \delta\mathbf{E}\Big[\max_{\eta(t)\in\mathscr{S}\in\mathscr{W}} V(x(t),\eta(t),t)\Big], \quad (12.7)$$

for all $t-2\chi \leq \xi \leq t$. Then the filter (12.2) is said to satisfy the prescribed $\mathscr{H}_\infty$ performance level $\gamma > 0$.

From here, we use (*) as an ellipsis for terms that are induced by symmetry in the symmetric block matrices. For notation simplicity, we will denote $\mathscr{A}_{ij}(\eta(t)) = \mathscr{A}_{ij}(\iota)$ when $\eta(t) = \iota \in \mathscr{S}$, and wherever appropriate.

12.2 Main Result

In this chapter, we assume that τ_k^s is bounded. For each $\eta(t) = \iota \in \mathscr{S}$, there is no loss of generality to assume $\tau(\iota,t) \leq \tau^*(\iota)$, where $\tau^*(\iota)$ is known positive constants. The following theorem provides sufficient conditions for the existence of a robust filter of the form (12.2) for the system (12.1) that satisfy $\mathscr{H}_\infty$ requirement.

Theorem 12.1. *Consider the system (12.1) satisfying Assumption 12.1. For given positive delay free attenuation constant γ_{d_f}, positive constants $\tau^*(\iota)$, ε_1, ε_2, ε_3, ε_4, ε_5, ε_6, and ε_7, if there exist symmetric positive matrices $X(\iota)$, $Y(\iota)$, R_{1_ι}, R_{2_ι}, R_{3_ι}, and R_{4_ι}, and matrices $F(\iota)$, and positive scalars β_{1_ι}, β_{2_ι}, such that the following inequalities (12.8)-(12.15) hold where $\iota \in \mathscr{S}$:*

$$\begin{bmatrix} Y(\iota) & I \\ I & X(\iota) \end{bmatrix} > 0, \quad (12.8)$$

$$\Upsilon_{ii}(\iota) < 0, \textit{ for } i \in \mathscr{I}_R \quad (12.9)$$

$$\Upsilon_{ij}(\iota) + \Upsilon_{ji}(\iota) < 0, \textit{ for } i < j < r \quad (12.10)$$

$$\Phi_{ij}(\iota) < 0, \textit{ for } \{i,j\} \in \mathscr{I}_R \times \mathscr{I}_R \quad (12.11)$$

$$\begin{bmatrix} R_{4_\iota} & (*)^T \\ \Lambda_\iota^T & \mathscr{Q}_\iota \end{bmatrix} > 0, \quad (12.12)$$

$$\begin{bmatrix} -R_{1_\iota} & (*)^T & (*)^T & (*)^T \\ 0 & -I & (*)^T & (*)^T \\ 0 & -Y(\iota) & -R_{2_\iota} & (*)^T \\ 0 & 0 & 0 & -R_{3_\iota} \end{bmatrix} < 0, \quad (12.13)$$

$$\begin{bmatrix} -\beta_{2_\iota} Y(\iota) & (*)^T & (*)^T & (*)^T & (*)^T & (*)^T \\ -\beta_{2_\iota} I & -\beta_{2_\iota} X(\iota) & (*)^T & (*)^T & (*)^T & (*)^T \\ 0 & Y(\iota) C_{2j}^T F_i^T(\iota) & -Y(\iota) & (*)^T & (*)^T & (*)^T \\ 0 & 0 & -I & -X(\iota) & (*)^T & (*)^T \\ 0 & \varepsilon_5 H_{3j}^T F_i^T(\iota) & 0 & 0 & -\varepsilon_5 I & (*)^T \\ 0 & 0 & E_{1j} Y(\iota) & 0 & 0 & -\varepsilon_5 I \end{bmatrix} < 0, \quad (12.14)$$

$$\begin{bmatrix} -Y(\iota) & (*)^T & (*)^T & (*)^T & (*)^T & (*)^T & (*)^T & (*)^T \\ -I & -X(\iota) & (*)^T & (*)^T & (*)^T & (*)^T & (*)^T & (*)^T \\ B_{1i}^T & B_{1i}^T X(\iota) & -I & (*)^T & (*)^T & (*)^T & (*)^T & (*)^T \\ 0 & D_{2j}^T F_i^T(\iota) & 0 & -I & (*)^T & (*)^T & (*)^T & (*)^T \\ \varepsilon_6 H_{1i}^T & \varepsilon_6 H_{1i}^T X(\iota) & 0 & 0 & -\varepsilon_6 I & (*)^T & (*)^T & (*)^T \\ 0 & 0 & E_{2i} & 0 & 0 & -\varepsilon_6 I & (*)^T & (*)^T \\ 0 & \varepsilon_7 H_{3j}^T F_i^T(\iota) & 0 & 0 & 0 & 0 & -\varepsilon_7 I & (*)^T \\ 0 & 0 & 0 & E_{2j} & 0 & 0 & 0 & -\varepsilon_7 I \end{bmatrix} < 0, \quad (12.15)$$

where $\Upsilon_{ij}(\iota)$ and $\Phi_{ij}(\iota)$ are expressed as:

$$\Upsilon_{ij}(\iota) = \begin{bmatrix} \begin{pmatrix} A_i Y(\iota) + Y(\iota) A_i^T \\ +(\beta_{1_\iota} + 4\beta_{2_\iota})\tau^*(\iota) Y(\iota) \\ +\lambda_{\iota\iota} Y(\iota) \end{pmatrix} & (*)^T & (*)^T \\ (\beta_{1_\iota} + 4\beta_{2_\iota})\tau^*(\iota) I & \begin{pmatrix} X(\iota)A + A^T X(\iota) \\ +F_i(\iota)C_{2j} + C_{2j}^T F_i^T(\iota) \\ +(\beta_{1_\iota} + 4\beta_{2_\iota})\tau^*(\iota) X(\iota) \\ +\sum_{\wp=1}^{s} \lambda_{\iota\wp} X(\wp) \end{pmatrix} & (*)^T \\ B_{1i}^T & B_{1i}^T X(\iota) & -\gamma_{d_f} I \\ 0 & D_{2j}^T F_{1i}^T(\iota) & 0 \\ C_{1i} Y(\iota) - L_i(\iota) & C_{1i} & 0 \\ E_{1i} Y(\iota) & E_{1i} & E_{2i} \\ \varepsilon_1 H_{1i}^T & \varepsilon_1 H_{1i}^T X(\iota) & 0 \\ E_{1j} Y(\iota) & E_{1j} & 0 \\ 0 & \varepsilon_2 H_{3j}^T F_i^T(\iota) & 0 \\ E_{1i} Y(\iota) & E_{1i} & 0 \\ 0 & 0 & 0 \\ S^T(\iota) & 0 & 0 \end{bmatrix}$$

$$\begin{bmatrix} (*)^T & (*)^T & (*)^T & (*)^T & (*)^T & (*)^T & (*)^T & (*)^T & (*)^T \\ (*)^T & (*)^T & (*)^T & (*)^T & (*)^T & (*)^T & (*)^T & (*)^T & (*)^T \\ (*)^T & (*)^T & (*)^T & (*)^T & (*)^T & (*)^T & (*)^T & (*)^T & (*)^T \\ -\gamma_{d_f} I & (*)^T & (*)^T & (*)^T & (*)^T & (*)^T & (*)^T & (*)^T & (*)^T \\ 0 & -I & (*)^T & (*)^T & (*)^T & (*)^T & (*)^T & (*)^T & (*)^T \\ 0 & 0 & -\varepsilon_1 I & (*)^T & (*)^T & (*)^T & (*)^T & (*)^T & (*)^T \\ 0 & 0 & 0 & -\varepsilon_1 I & (*)^T & (*)^T & (*)^T & (*)^T & (*)^T \\ E_{2j} & 0 & 0 & 0 & -\varepsilon_2 I & (*)^T & (*)^T & (*)^T & (*)^T \\ 0 & 0 & 0 & 0 & 0 & -\varepsilon_2 I & (*)^T & (*)^T & (*)^T \\ 0 & 0 & 0 & 0 & 0 & 0 & -\varepsilon_3 I & (*)^T & (*)^T \\ 0 & \varepsilon_3 H_{2i}^T & 0 & 0 & 0 & 0 & 0 & -\varepsilon_3 I & (*)^T \\ 0 & 0 & 0 & 0 & 0 & 0 & 0 & 0 & -\mathscr{Q}_\iota \end{bmatrix},$$

$$\Phi_{ij}(\iota) = \begin{bmatrix} -\beta_{1_\iota}Y(\iota)+R_{1_\iota} & (*)^T & (*)^T & (*)^T & (*)^T & (*)^T & (*)^T \\ -\beta_{1_\iota}I & -\beta_{1_\iota}X(\iota) & (*)^T & (*)^T & (*)^T & (*)^T & (*)^T \\ Y(\iota)A_i^T & -A_i - Y(\iota)C_{2j}^T F_i^T(\iota) - \lambda_{\iota\iota}I & -Y(\iota)+R_{2_\iota} & (*)^T & (*)^T & (*)^T & (*)^T \\ A_i^T & A_i^T X(\iota) & -I & -X(\iota)+R_{3_\iota} & (*)^T & (*)^T & (*)^T \\ \varepsilon_3 H_{1i}^T & \varepsilon_3 H_{1i}^T X(\iota) & 0 & 0 & -\varepsilon_4 I & (*)^T & (*)^T \\ 0 & 0 & E_{1i}Y(\iota) & E_{1i} & 0 & -\varepsilon_4 I & (*)^T \\ 0 & R_{4_\iota} & 0 & 0 & 0 & 0 & -I \end{bmatrix},$$

and

$$S(\iota) = [\sqrt{\lambda_{\iota 1}}Y(\iota) \cdots \sqrt{\lambda_{\iota(\iota-1)}}Y(\iota)\sqrt{\lambda_{\iota(\iota+1)}}Y(\iota) \cdots \sqrt{\lambda_{\iota s}}Y(\iota)],$$
$$\Lambda_\iota = [\sqrt{\lambda_{\iota 1}}I \cdots \sqrt{\lambda_{\iota(\iota-1)}}I \; \sqrt{\lambda_{\iota(\iota+1)}}I \cdots \sqrt{\lambda_{\iota s}}I],$$

with

$$\mathcal{Q}_\iota = diag\{Y(1,k), \cdots, Y(\iota-1,k), Y(\iota+1,k), \cdots, Y(s,k)\},$$

then (12.6) holds for all delays $\tau(\iota,t)$ *satisfying* $\tau(\iota,t) \le \tau^*(\iota)$ *with* $\gamma^2 = \gamma_{d_f} + \max(\tau^*(\iota))$. *Furthermore, the mode dependant robust filter* $\mathcal{F}$ *is of the form (12.2) with*

$$\hat{A}_{ij}(\iota) = [Y^{-1}(\iota) - X(\iota)]^{-1}[-A_i^T - X(\iota)A_iY(\iota)$$
$$-F_i(\iota)C_{2j}Y(\iota) - \sum_{\wp=1}^{s} \lambda_{\iota\wp}Y^{-1}(\wp)Y(\iota)]Y^{-1}(\iota), \tag{12.16}$$
$$\hat{B}_i(\iota) = [Y^{-1}(\iota) - X(\iota)]^{-1}F_i(\iota), \tag{12.17}$$
$$\hat{C}_i(\iota) = L_i(\iota)Y^{-1}(\iota). \tag{12.18}$$

Proof. The results can be obtained employing the same technique used in Chapter 6 and Chapter 9. It is basically an application of Lyapunov stability theorem and Razumikhin-type theorem [105] for stochastic systems with Markovian jumps in the realm of fuzzy logic control. □

The iterative algorithm presented in Chapter 9 is applied here so solve Theorem 12.1 which is a BMI problem.

12.3 Example

To illustrate the validation of the results obtained in this chapter, we consider the same plant as in previous chapter. The fuzzy rules and the transition rate matrices remain unchanged.

In our simulation, we assume $\tau^*(1) = 0.03$ and $\tau^*(2) = 0.05$. We assume the sampling period is 0.015, that is, $h^s = 0.015$, and $n^s = 0$ which means no data packet dropout happens in the communication channel. Delay free attenuation constant γ_{d_f} is set to be 1, while constants ε_1, ε_2, ε_3, ε_4, ε_5, ε_6, and ε_7 are set be equivalent to 1.

In the following simulation, we assume $F(t) = \sin t$ and it can be seen that $\|F(t)\| \leq 1$.

By applying Theorem 12.1 and the algorithm in the previous section, we get the following controller gains by the calculation of (12.16)-(12.18):

$$\hat{A}_{11}(1) = \begin{bmatrix} 10.53 & 2.0092 \\ -34.0107 & -12.2455 \end{bmatrix}, \hat{A}_{12}(1) = \begin{bmatrix} 10.4366 & 2.0933 \\ -34.9968 & -12.9857 \end{bmatrix},$$
$$\hat{A}_{21}(1) = \begin{bmatrix} 10.3821 & 2.1759 \\ -34.0224 & -12.1141 \end{bmatrix}, \hat{A}_{22}(1) = \begin{bmatrix} 10.6285 & 2.1346 \\ -34.2610 & -12.3825 \end{bmatrix},$$
$$\hat{B}_1(1) = \begin{bmatrix} 12.8827 \\ -2.8821 \end{bmatrix}, \hat{B}_2(1) = \begin{bmatrix} 12.9087 \\ -2.7922 \end{bmatrix},$$
$$\hat{C}_1(1) = \begin{bmatrix} -0.8976 & -6.7611 \end{bmatrix}, \hat{C}_2(1) = \begin{bmatrix} -0.8812 & -6.6696 \end{bmatrix},$$
$$\hat{A}_{11}(2) = \begin{bmatrix} -43.5656 & -19.8822 \\ 107.8762 & -29.0972 \end{bmatrix}, \hat{A}_{12}(2) = \begin{bmatrix} -43.6012 & -19.9212 \\ 109.2281 & -29.1222 \end{bmatrix},$$
$$\hat{A}_{21}(2) = \begin{bmatrix} -43.3244 & -19.2342 \\ 109.1202 & -28.3822 \end{bmatrix}, \hat{A}_{22}(2) = \begin{bmatrix} -43.1233 & -19.1235 \\ 107.9383 & -28.8433 \end{bmatrix},$$
$$\hat{B}_1(2) = \begin{bmatrix} 11.2983 \\ -7.3432 \end{bmatrix}, \hat{B}_2(2) = \begin{bmatrix} 11.0211 \\ -7.2331 \end{bmatrix},$$
$$\hat{C}_1(2) = \begin{bmatrix} 5.8923 & -3.3934 \end{bmatrix}, \hat{C}_2(2) = \begin{bmatrix} 5.9033 & -3.4077 \end{bmatrix},$$

The ratio of the filer error energy to the disturbance input noise is depicted in Figure 12.2. In our simulation, we use a uniform distributed random disturbance input signal $w(t)$ with maximum value 1.5. It can be seen that the ratio tends to a constant value of about 2.1×10^{-5}, which means the attenuation level equals to $\sqrt{2.1 \times 10^{-5}} \approx 4.6 \times 10^{-3}$, less than the prescribed level $\gamma = \sqrt{\gamma_{d_f} + \max(\tau^*(\iota))} = \sqrt{1 + 0.03} \approx 1.01$.

In conclusion, the designed filter meets the performance requirements.

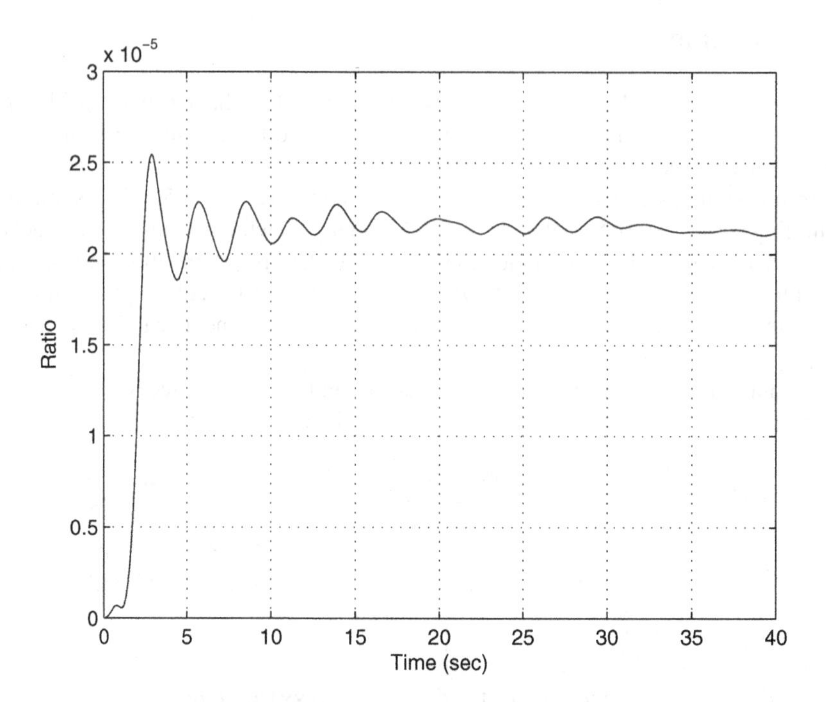

Fig. 12.2 The ratio of the filter error energy to the disturbance input noise without data dropouts

12.4 Conclusion

In this chapter, a technique of designing a delay-dependant robust fuzzy filter for an uncertain nonlinear NCS with random communication network-induced delays and data packet dropouts has been proposed. The main contribution of this work is that network-induced effects are regarded as input delays, and the designed filter is mode-dependant by using Markov process to model the network-induced effects, which reduces the conservatism effectively. The Lyapunov–Razumikhin method has been employed to derive such a robust filter for this class of systems. Sufficient conditions for the existence of such a filter for this class of NCSs are derived.

Chapter 13
Fault Estimation for Uncertain Nonlinear Networked Control Systems

This chapter proposes a robust fuzzy fault estimator for a class of nonlinear uncertain NCSs that ensures the fault estimation error is less than prescribed $\mathscr{H}_\infty$ performance level, irrespective of the uncertainties and network-induced effects. Sufficient conditions for the existence of such a fault estimator for this class of NCSs are derived in terms of the solvability of BMIs.

13.1 Problem Formulation and Preliminaries

In this chapter, we describe the nonlinear NCSs as follows:

$$\begin{cases} \dot{x}(t) = \sum_{i=1}^{r} \mu_i(\nu(t))[(A_i + \Delta A_i)x(t) + B_i w(t) + G_i f(t)] \\ y(t) = \sum_{i=1}^{r} \mu_i(\nu(t))[(C_i + \Delta C_i)x(t) + D_i w(t) + J_i f(t)] \end{cases} \tag{13.1}$$

where $x(t) \in \mathbb{R}^n$ is the state vector, $w(t) \in \mathbb{R}^p$ and $f(t) \in \mathbb{R}^q$ are, respectively, exogenous disturbances and faults which belong to $\mathscr{L}_2[0,\infty)$, $y(t) \in \mathbb{R}^l$ denotes the measurement output.

Furthermore, $i \in \mathscr{I}_R = \{1,\cdots,r\}$, r is the number of fuzzy rules; $\nu_k(t)$ are premise variables, $M_{\iota\kappa}$ are fuzzy sets, $k = 1,\cdots,p$, p is the number of premise variables

$$\nu(t) = [\nu_1(t), \nu_2(t), \cdots, \nu_p(t)]^T,$$

$$\omega_i(\nu(t)) = \prod_{k=1}^{p} M_{\iota\kappa}(\nu_k(t)),\ \omega_i(\nu(t)) \geq 0,\ \sum_{i=1}^{r} \omega_i(\nu(t)) > 0,$$

$$\mu_i(\nu(t)) = \frac{\omega_i(\nu(t))}{\sum_{i=1}^{r} \omega_i(\nu(t))},\ \mu_i(\nu(t)) \geq 0,\ \sum_{i=1}^{r} \mu_i(\nu(t)) = 1.$$

Here, $M_{\iota\kappa}(\nu_k(t))$ denote the grade of membership of $\nu_k(t)$ in $M_{\iota\kappa}$.

In addition, matrices ΔA_i and ΔC_i characterize the uncertainties in the system and satisfy the following assumption:

D. Huang and S.K. Nguang: Robust Ctrl. for Uncertain Networked Ctrl. Sys., LNCIS 386, pp. 137–144.
springerlink.com

Assumption 13.1.

$$\begin{bmatrix} \Delta A_i \\ \Delta C_i \end{bmatrix} = \begin{bmatrix} H_{1i} \\ H_{2i} \end{bmatrix} F(t)E_i,$$

where H_{1i}, H_{2i}, and E_i are known real constant matrices of appropriate dimensions, and $F(t)$ is an unknown matrix function with Lebesgue-measurable elements and satisfies $F(t)^T F(t) \leq I$, in which I is the identity matrix of appropriate dimension.

In this chapter, we consider a nonlinear NCS of which the plant is described by the T-S model (13.1). The setup of the overall configuration is depicted in Figure 13.1, where $\tau(t) \geq 0$ is the random time-delay from sensor to controller. These delays are assumed to be upper bounded. Furthermore, we apply the same Markov processes introduced in Chapter 2 to model the random time-delays in this chapter. In the

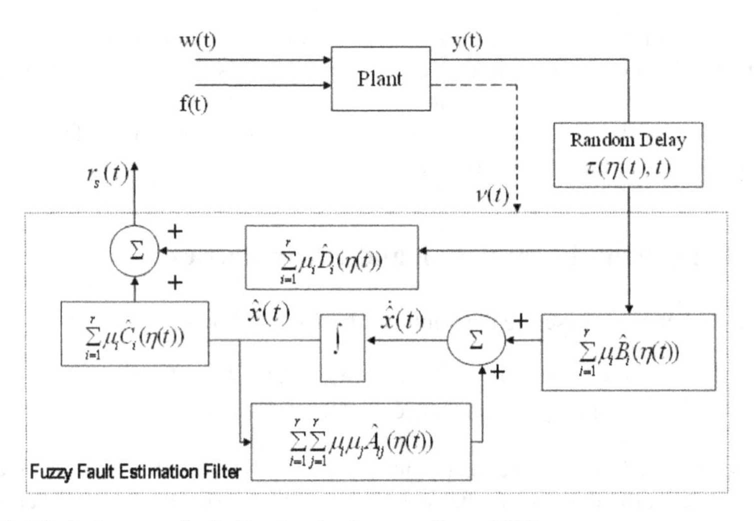

Fig. 13.1 Block diagram of a fault estimator for a nonlinear NCS

system setup, the premise vector $v(t)$ is connected to the fault estimator via point-to-point architecture, which is immune to network-induced delays.

Therefore, following the modeling procedure presented in Chapter 2, for the nonlinear plant represented by (13.1), the fuzzy dynamic output feedback controller at time t is inferred as follows:

$$\begin{aligned} \dot{\hat{x}}(t) &= \textstyle\sum_{i=1}^{r} \sum_{j=1}^{r} \mu_i(v(t))\mu_j(v(t)) \left[\hat{A}_{ij}(\eta(t))\hat{x}(t) + \hat{B}_i(\eta(t))y(t - \tau(\eta(t),t))\right] \\ u(t) &= \textstyle\sum_{i=1}^{r} \mu_i(v(t)) \left[\hat{C}_i(\eta(t))\hat{x}(t) + \hat{D}_i(\eta(t))y(t - \tau(\eta(t),t))\right], \end{aligned} \tag{13.2}$$

where $\hat{A}_{ij}(\eta(t))$, $\hat{B}_i(\eta(t))$, $\hat{C}_i(\eta(t))$ and $\hat{D}_i(\eta(t))$ in each plant rule are parameters of the fault estimator which are to be designed.

Substituting (13.2) into (13.1) yields

$$\begin{aligned}\dot{\tilde{x}}(t) &= \sum_{i=1}^{r}\sum_{j=1}^{r}\mu_i(v(t))\mu_j(v(t))\Big[\mathscr{A}_{ij}(\eta(t))\tilde{x}(t)+\mathscr{B}_{ij}(\eta(t))\tilde{x}(t-\tau(\eta(t),t))\\&\quad+\mathscr{C}_{ij}(\eta(t))\omega(t)\Big]\\e(t) &= \sum_{i=1}^{r}\sum_{j=1}^{r}\mu_i(v(t))\mu_j(v(t))\Big[\mathscr{D}_{1i}(\eta(t))\tilde{x}(t)+\mathscr{D}_{2ij}(\eta(t))\tilde{x}(t-\tau(\eta(t),t))\\&\quad+\mathscr{D}_{3ij}(\eta(t))\omega(t)\Big],\end{aligned}\tag{13.3}$$

where $e(t)=r_s(t)-f(t)$ is the fault estimation error, $\omega(t)=[w^T(t)\;\; f^T(t)\;\; w^T(t-\tau(\eta(t),t))\;\; f^T(t-\tau(\eta(t),t))]^T$, $\tilde{x}(t)=[x^T(t)\;\; \hat{x}^T(t)]^T$, and

$$\begin{aligned}&\mathscr{A}_{ij}(\eta(t))=\begin{bmatrix}A_i+\Delta A_i & 0\\ 0 & \hat{A}_{ij}(\eta(t))\end{bmatrix},\ \mathscr{B}_{ij}(\eta(t))=\begin{bmatrix}0 & 0\\ \hat{B}_i(\eta(t))(C_j+\Delta C_j) & 0\end{bmatrix},\\&\mathscr{C}_{ij}(\eta(t))=\begin{bmatrix}B_i & G_i & 0 & 0\\ 0 & 0 & \hat{B}_i(\eta(t))D_j & \hat{B}_i(\eta(t))J_j\end{bmatrix},\\&\mathscr{D}_{1i}(\eta(t))=\begin{bmatrix}0 & \hat{C}_i(\eta(t))\end{bmatrix},\ \mathscr{D}_{2ij}(\eta(t))=\begin{bmatrix}\hat{D}_i(\eta(t))(C_j+\Delta C_j) & 0\end{bmatrix},\\&\mathscr{D}_{3ij}(\eta(t))=\begin{bmatrix}0 & -I & \hat{D}_i(\eta(t))D_j & \hat{D}_i(\eta(t))J_j\end{bmatrix}.\end{aligned}$$

The aim of this chapter is to design a fault estimator of the form (13.2) such that the following inequality holds:

For (13.3) with its zero state response ($x(\phi)=0,\ \omega(\phi)=0,\ -\chi\le\phi\le 0$),

$$\mathbf{E}\Big[\int_0^{T_f} e^T(t)e(t)dt\Big]\le\gamma^2\mathbf{E}\Big[\int_0^{T_f}\sup_{-\chi\le\phi\le 0}\omega^T(t+\phi)\omega(t+\phi)dt\Big],\tag{13.4}$$

for any nonzero $\omega(t)\in\mathscr{L}_2[0,\ T_f]$ and $T_f\ge 0$, provided $x=\{x(\xi): t-2\chi\le\xi\le t\}\in L^2_{\mathscr{F}_t}([-2\chi,0];\mathbb{R}^n)$ satisfying:

$$\mathbf{E}\Big[\min_{\eta(t)\in\mathscr{S}\in\mathscr{W}}V(x(\xi),\eta(\xi))\Big]<\delta\mathbf{E}\Big[\max_{\eta(t)\in\mathscr{S}\in\mathscr{W}}V(x(t),\eta(t),t)\Big],\tag{13.5}$$

for all $t-2\chi\le\xi\le t$, then a fault estimator is designed satisfying a disturbance attenuation level γ.

In this chapter, we assume $u(t)=0$ before the first control signal reaches the plant. From here, $\mu_i(v(t))$ and $\mu_j(v(t))$ are denoted as μ_i and μ_j respectively for the convenience of notations. In the symmetric block matrices, we use (*) as an ellipsis for terms that are induced by symmetry. $\hat{A}_{ij}(\eta(t))$ is denoted as $\hat{A}_{ij}(\iota)$ if $\eta(t)=\iota$.

13.2 Main Result

The following theorem provides sufficient conditions for the existence of a mode-dependent fault estimator for the system (13.3) that guarantees disturbance attenuation level γ.

Theorem 13.1. *Consider the system (13.3) satisfying Assumption 13.1. For given positive delay-free attenuation constant γ_{d_f}, positive constants $\tau^*(i)$, ε_{1ij_ι}, ε_{2ij_ι}, ε_{3ij_ι}, and ε_{4ij_ι}, if there exist symmetric positive matrices $X(\iota)$, $Y(\iota)$, R_{1_ι}, R_{2_ι}, R_{3_ι}, and R_{4_ι}, and matrices $F_i(\iota)$, $L_i(\iota)$, and $\hat{D}_i(\iota)$, and positive scalars β_{1_ι}, β_{2_ι}, such that the following inequalities hold where $\iota \in \mathscr{S}$:*

$$\begin{bmatrix} Y(\iota) & I \\ I & X(\iota) \end{bmatrix} > 0, \tag{13.6}$$

$$\Upsilon_{ii}(\iota) < 0, \; for \; i \in \mathscr{I}_R \tag{13.7}$$

$$\Upsilon_{ij}(\iota) + \Upsilon_{ji}(\iota) < 0, \; for \; i < j < r \tag{13.8}$$

$$\Phi_{ij}(\iota) < 0, \; for \; \{i,j\} \in \mathscr{I}_R \times \mathscr{I}_R \tag{13.9}$$

$$\begin{bmatrix} R_{4_\iota} & (*)^T \\ \Lambda_\iota^T & \mathscr{Q}(\iota) \end{bmatrix} > 0, \tag{13.10}$$

$$\begin{bmatrix} -R_{1_\iota} & (*)^T & (*)^T & (*)^T \\ 0 & -I & (*)^T & (*)^T \\ 0 & -Y(\iota) & -R_{2_\iota} & (*)^T \\ 0 & 0 & 0 & -R_{3_\iota} \end{bmatrix} < 0, \tag{13.11}$$

$$\begin{bmatrix} -\beta_{2_\iota} Y(\iota) & (*)^T & (*)^T & (*)^T & (*)^T & (*)^T \\ -\beta_{2_\iota} I & -\beta_{2_\iota} X(\iota) & (*)^T & (*)^T & (*)^T & (*)^T \\ 0 & Y(\iota) C_j^T F_i^T(\iota) & -Y(\iota) & (*)^T & (*)^T & (*)^T \\ 0 & C_j^T F_i^T(\iota) & -I & -X(\iota) & (*)^T & (*)^T \\ 0 & \varepsilon_{4ij_\iota} H_{2j}^T F_i^T(\iota) & 0 & 0 & -\varepsilon_{4ij_\iota} I & (*)^T \\ 0 & 0 & E_j Y(\iota) & E_j & 0 & -\varepsilon_{4ij_\iota} I \end{bmatrix} < 0,$$

$$for \; \{i,j\} \in \mathscr{I}_R \times \mathscr{I}_R \tag{13.12}$$

$$\begin{bmatrix} -Y(\iota) & (*)^T & (*)^T & (*)^T & (*)^T & (*)^T \\ -I & -X(\iota) & (*)^T & (*)^T & (*)^T & (*)^T \\ B_i^T & B_i^T X(\iota) & -I & (*)^T & (*)^T & (*)^T \\ G_i^T & G_i^T X(\iota) & 0 & -I & (*)^T & (*)^T \\ 0 & D_i^T F_i^T(\iota) & 0 & 0 & -I & (*)^T \\ 0 & J_j^T F_i^T(\iota) & 0 & 0 & 0 & -I \end{bmatrix} < 0,$$

$$for \; \{i,j\} \in \mathscr{I}_R \times \mathscr{I}_R \tag{13.13}$$

where

$$
\Phi_{ij}(\iota)=\begin{bmatrix}
-\beta_{1_\iota}Y(\iota)+R_{1_\iota} & (*)^T & (*)^T & (*)^T & (*)^T & (*)^T & (*)^T\\
-\beta_{1_\iota}I & -\beta_{1_\iota}X(\iota) & (*)^T & (*)^T & (*)^T & (*)^T & (*)^T\\
Y(\iota)A_i^T & -A_i-Y(\iota)C_j^TF_i^T(\iota)-\lambda_{\iota\iota}I & -Y(\iota)+R_{2_\iota} & (*)^T & (*)^T & (*)^T & (*)^T\\
A_i^T & A_i^TX(\iota) & -I & -X(\iota)+R_{3_\iota} & (*)^T & (*)^T & (*)^T\\
\varepsilon_{3ij_\iota}H_{1i}^T & \varepsilon_{3ij_\iota}H_{1i}^TX(\iota) & 0 & 0 & -\varepsilon_{3ij_\iota}I & (*)^T & (*)^T\\
0 & 0 & E_iY(\iota) & E_i & 0 & -\varepsilon_{3ij_\iota}I & (*)^T\\
0 & R_{4_\iota} & 0 & 0 & 0 & 0 & -I
\end{bmatrix}
$$

$$
\Upsilon_{ij}(\iota)=\begin{bmatrix}
\Xi_{1i}(\iota) & (*)^T & (*)^T & (*)^T & (*)^T & (*)^T & (*)^T & (*)^T & (*)^T & (*)^T & (*)^T & (*)^T & (*)^T & (*)^T\\
(\beta_{1_\iota}+4\beta_{2_\iota})\tau^*(\iota)I & \Xi_{2ij}(\iota) & (*)^T & (*)^T & (*)^T & (*)^T & (*)^T & (*)^T & (*)^T & (*)^T & (*)^T & (*)^T & (*)^T & (*)^T\\
0 & 0 & -I & (*)^T & (*)^T & (*)^T & (*)^T & (*)^T & (*)^T & (*)^T & (*)^T & (*)^T & (*)^T & (*)^T\\
0 & 0 & 0 & -I & (*)^T & (*)^T & (*)^T & (*)^T & (*)^T & (*)^T & (*)^T & (*)^T & (*)^T & (*)^T\\
B_i^T & B_i^TX(\iota) & 0 & 0 & -\gamma_{d_f}I & (*)^T & (*)^T & (*)^T & (*)^T & (*)^T & (*)^T & (*)^T & (*)^T & (*)^T\\
G_i^T & G_i^TX(\iota) & 0 & 0 & 0 & -\gamma_{d_f}I & (*)^T & (*)^T & (*)^T & (*)^T & (*)^T & (*)^T & (*)^T & (*)^T\\
0 & D_j^TF_i(\iota) & 0 & 0 & 0 & 0 & -\gamma_{d_f}I & (*)^T & (*)^T & (*)^T & (*)^T & (*)^T & (*)^T & (*)^T\\
0 & J_j^TF_i(\iota) & 0 & 0 & 0 & 0 & 0 & -\gamma_{d_f}I & (*)^T & (*)^T & (*)^T & (*)^T & (*)^T & (*)^T\\
L(\iota) & 0 & \hat{D}_i(\iota)C_j & 0 & 0 & -I & \hat{D}_i(\iota)D_j & \hat{D}_i(\iota)J_j & -I & (*)^T & (*)^T & (*)^T & (*)^T & (*)^T\\
Z^T(\iota) & 0 & 0 & 0 & 0 & 0 & 0 & 0 & 0 & -\mathcal{Q}(\iota) & (*)^T & (*)^T & (*)^T & (*)^T\\
\varepsilon_{1ij_\iota}H_{1i}^T & \varepsilon_{1ij_\iota}(H_{1i}^TX^T(\iota)+H_{2j}^TF_i^T(\iota)) & 0 & 0 & 0 & 0 & 0 & 0 & 0 & 0 & -\varepsilon_{1ij_\iota}I & (*)^T & (*)^T & (*)^T\\
E_iY(\iota) & E_i & 0 & 0 & 0 & 0 & 0 & 0 & 0 & 0 & 0 & -\varepsilon_{1ij_\iota}I & (*)^T & (*)^T\\
0 & 0 & 0 & 0 & 0 & 0 & 0 & 0 & \varepsilon_{2ij_\iota}H_{2j}^T\hat{D}_i^T(\iota) & 0 & 0 & 0 & -\varepsilon_{2ij_\iota}I & (*)^T\\
0 & 0 & E_i & 0 & 0 & 0 & 0 & 0 & 0 & 0 & 0 & 0 & 0 & -\varepsilon_{2ij_\iota}I
\end{bmatrix}
$$

and

$$
\begin{aligned}
Z(\iota) &= [\sqrt{\lambda_{\iota 1}}Y(\iota)\cdots\sqrt{\lambda_{\iota(\iota-1)}}Y(\iota)\ \sqrt{\lambda_{\iota(\iota+1)}}Y(\iota)\cdots\sqrt{\lambda_{\iota s}}Y(\iota)],\\
\Lambda_\iota &= [\sqrt{\lambda_{\iota 1}}I\cdots\sqrt{\lambda_{\iota(\iota-1)}}I\ \sqrt{\lambda_{\iota(\iota+1)}}I\cdots\sqrt{\lambda_{\iota s}}I],\\
\mathscr{Q}(\iota) &= diag\{Y(1),\cdots,Y(\iota-1),Y(\iota+1),\cdots,Y(s)\},\\
\Xi_{1i}(\iota) &= A_iY(\iota)+Y(\iota)A_i^T+(\beta_{1_\iota}+4\beta_{2_\iota})\tau^*(\iota)Y(\iota)+\lambda_{\iota\iota}Y(\iota),\\
\Xi_{2ij}(\iota) &= X(\iota)A_i+A_i^TX(\iota)+F_i(\iota)C_j+C_j^TF_i^T(\iota)\\
&\quad+(\beta_{1_\iota}+4\beta_{2_\iota})\tau^*(\iota)X(\iota)+\sum_{\wp=1}^{s}\lambda_{\iota\wp}X(\wp),
\end{aligned}
$$

then (13.3) holds for delay $\tau(\iota,t)$ satisfying $\tau(\iota,t)\le\tau^*(\iota)$ with $\gamma^2=\gamma_{d_f}+\max(\tau^*(\iota))$ for $\iota\in\mathscr{S}$. Furthermore, the mode dependant fault estimator is obtained of the form (13.2) with

$$
\begin{aligned}
\hat{A}_{ij}(\iota) &= [Y^{-1}(\iota)-X(\iota)]^{-1}[-A_i^T-X(\iota)A_iY(\iota)-F_i(\iota)C_jY(\iota)\\
&\quad-\sum_{\wp=1}^{s}\lambda_{\iota\wp}Y^{-1}(\wp)Y(\iota)]Y^{-1}(\iota), \qquad (13.14)\\
\hat{B}_i(\iota) &= [Y^{-1}(\iota)-X(\iota)]^{-1}F_i(\iota), \qquad (13.15)\\
\hat{C}_i(\iota) &= L_i(\iota)Y^{-1}(\iota). \qquad (13.16)
\end{aligned}
$$

Proof. It is straightforward from the proof process in Chapter 9 and Chapter 7. □

The iterative algorithm presented in Chapter 9 is applied here so solve Theorem 13.1 which is a BMI problem.

13.3 Numerical Example

To illustrate the validation of the results obtained in this chapter, we consider the same plant as set in Chapter 11. The fuzzy rules and the transition rate matrices remain unchanged. In our simulation, we assume $\tau^*(1)=0.045$ and $\tau^*(2)=0.025$. We assume the sampling period is 0.01, that is, $h^s=0.01$, and $n^s=0$ which means no data packet dropout happens in the communication channel.

The random time-delays exist in $\mathscr{S}=\{1,2\}$, and its transition rate matrices are given by:

$$
\Lambda=\begin{bmatrix}-1 & 1\\ 2 & -2\end{bmatrix}.
$$

In this example, the fault signal is simulated as follows:

$$
f(t)=\begin{cases}1 & t\in[5,\ 10]\\ 0 & others.\end{cases} \qquad (13.17)
$$

For the sake of simplicity, $\hat{D}_i$ is assumed to be a zero matrix in this example. By applying Theorem 13.1 and the iterative algorithm, we get the following fault estimator for $i \in \mathscr{S} = \{1,2\}$ of the form (13.14)-(13.16) where:

$$
\begin{aligned}
\hat{A}_{11}(1) &= \begin{bmatrix} -5.2761 & -42.358 \\ 79.949 & -18.168 \end{bmatrix}, \hat{A}_{12}(1) = \begin{bmatrix} -6.3274 & -41.749 \\ 82.695 & -18.11 \end{bmatrix}, \\
\hat{A}_{21}(1) &= \begin{bmatrix} -6.1284 & -44.1547 \\ 74.265 & -19.541 \end{bmatrix}, \hat{A}_{22}(1) = \begin{bmatrix} -6.1147 & -43.224 \\ 78.4474 & -19.3218 \end{bmatrix}, \\
\hat{B}_1(1) &= \begin{bmatrix} 1.0018 \\ 0.29931 \end{bmatrix}, \hat{B}_2(1) = \begin{bmatrix} 0.91953 \\ 0.18201 \end{bmatrix}, \\
\hat{C}_1(1) &= \begin{bmatrix} 2.06 & -7.8782 \end{bmatrix}, \hat{C}_2(1) = \begin{bmatrix} 1.9243 & -7.6107 \end{bmatrix}, \\
\hat{A}_{11}(2) &= \begin{bmatrix} -10.386 & -41.1 \\ 96.295 & -17.874 \end{bmatrix}, \hat{A}_{12}(2) = \begin{bmatrix} -2.4862 & -42.937 \\ 70.708 & -18.469 \end{bmatrix}, \\
\hat{A}_{21}(2) &= \begin{bmatrix} -8.546 & -44.587 \\ 85.4447 & -17.214 \end{bmatrix}, \hat{A}_{22}(2) = \begin{bmatrix} -2.5548 & -45.254 \\ 88.214 & -17.228 \end{bmatrix}, \\
\hat{B}_1(2) &= \begin{bmatrix} 1.1029 \\ -0.27829 \end{bmatrix}, \hat{B}_2(2) = \begin{bmatrix} 1.0011 \\ -0.0094 \end{bmatrix}, \\
\hat{C}_1(2) &= \begin{bmatrix} 2.0784 & -8.0286 \end{bmatrix}, \hat{C}_2(2) = \begin{bmatrix} 1.9683 & -7.7673 \end{bmatrix}.
\end{aligned}
$$

Histories of the residual signals $r_s(t)$ along with the fault signal $f(t)$ are shown in Figure 13.2. The results demonstrate that the designed fault estimator meets the performance requirement.

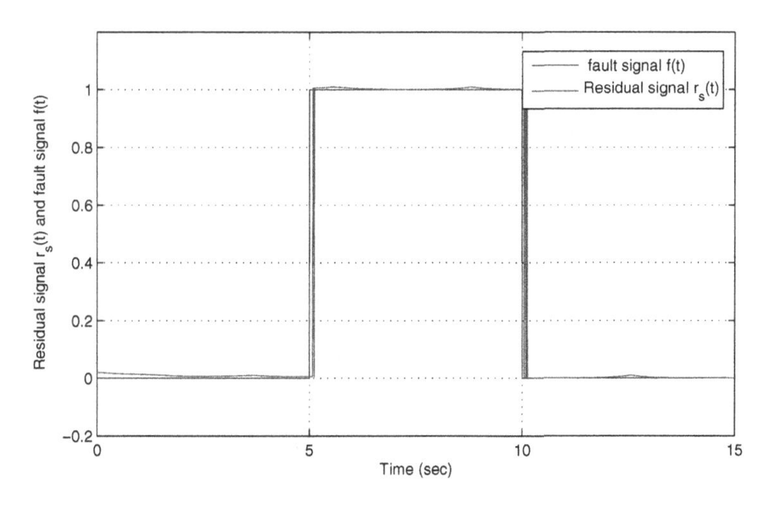

Fig. 13.2 Residual signals $r_s(t)$ and $f(t)$

13.4 Conclusion

In this chapter, a technique of designing a delay-dependant fuzzy fault estimator for an uncertain NCS with random communication network-induced delays and data packet dropouts has been proposed. The Lyapunov–Razumikhin method has been employed to derive such a controller for this class of systems. Sufficient conditions for the existence of such a controller for this class of NCSs are derived in a form of BMIs. We finally use a numerical example to demonstrate the effectiveness of this methodology at the last section.

Chapter 14
Conclusions

14.1 Summary of the Book

This book proposes novel methodologies for stability analysis, disturbance attenuation, and fault estimation for a class of linear/nonlinear uncertain NCSs with random communication network-induced delays and data packet dropouts in both sensor-to-controller and controller-to-actuator channels. Models for such network-induced effects are first developed by using Markov processes. Based on the Lyapunov–Razumikhin method, the existence of the designed controllers and fault estimators are given in terms of the solvability of BMIs. Iterative algorithms are proposed to change this non-convex problem into quasi-convex optimization problems, which can be solved effectively by available mathematical tools. The effectiveness and advantages of the proposed design methodologies are verified by numerical examples in each chapter. The simulation results show that the proposed design methodologies can achieve the prescribed performance requirement.

To clarify this approach, this book are divided into two parts. Part I is focused on the linear uncertain NCSs, while Part II is concentrated on the nonlinear uncertain NCSs. Chapter 2 is given to detail the modeling procedure of NCSs used in this book.

In Part I, Chapters 3 and 4 presents the synthesis design procedure of a robust state feedback controller and a robust dynamic output feedback controller, respectively. Chapters 5 presents the synthesis design procedure of a robust delay-dependant controller guarantees both robust stability and a prescribed disturbance attenuation performance for the closed-loop NCSs. Lastly, Chapter 6 gives the design procedure of a robust fault estimator that ensures the fault estimation error is less than a prescribed performance level, irrespective of the uncertainties and network-induced effects.

In Part II, preliminary background knowledge of the T-S fuzzy model, which is an ideal of model of nonlinear systems, is first given in Chapter 7. Then a robust state feedback controller and a robust dynamic output feedback controller for stochastic stability are given respectively in Chapter 8 and Chapter 9. In Chapter 10, we further develop an output feedback controller so that both robust stability and a prescribed

D. Huang and S.K. Nguang: Robust Ctrl. for Uncertain Networked Ctrl. Sys., LNCIS 386, pp. 145–146.
springerlink.com

disturbance attenuation performance for the closed-loop NCSs are achieved. A fault estimator for nonlinear NCSs is lastly given in Chapter 11.

Here is a summary of the contributions of this book:

- Network-induced delays and data packet dropouts in both the sensor-controller and controller-actuator channels are considered.
- Markovian processes are used to model the network-induced delays and data packet dropouts.
- T-S fuzzy model is used in the controller design for nonlinear NCSs; special care has been taken for the premise variables for the fuzzy rules.

As a result, this book provides an integrated approach for the design of NCSs and represents a valuable and meaningful contribution to the development of an LMI-based NCSs.

14.2 Future Research Work

In general, NCSs still remain to be an open area that lots of research works are required. Further research work, to name a few, could be directed to the following areas:

1. Another interesting consideration for future research work is involving the design of a mixed $\mathscr{H}_2/\mathscr{H}_\infty$ controller that can be lead into the optimal performance for nominal plant and guarantees stability against dynamic uncertainty. In other words, a mixed $\mathscr{H}_2/\mathscr{H}_\infty$ problem providing substantial amount of progress toward both robustness and multi-objective $\mathscr{H}_2$ problems.
2. Measurement quantization effect, which is one of essential issues in the implementation of NCSs. This quantization effect produces a further layer of complexity to the NCSs problems and needs further research attentions.
3. The properties of the controller with timeout are not completely studied.
4. This book assumes the state of the Markov chain is known. It will be an interesting problem when the state of the Markov chain is unknown. This problem needs further investigation. Furthermore, it would also be interesting to extend the results of this book to more general stochastic systems, such as systems with non-Markovian jumps.
5. Other types of uncertainties structures may be considered to cover more general systems.

Appendix A
Mathematical and Background Knowledge

In this chapter, we will introduce some mathematical background knowledge that will be applied throughout this research. Some essential matrix inequalities and other lemmas used in this research will also be presented.

A.1 Linear Matrix Inequality

Since the early 1990s, with the development of interior-point methods for solving LMI problems, the LMI method [98] has gained increased interest and emerged as useful tools for solving a number of control problems, such as synthesis of gain-scheduled (parameter-varying) controllers, mixed-norm and multi-objective control design, hybrid dynamical systems, and fuzzy control. Three factors make LMI techniques appealing:

- A variety of design specifications and constraints can be expressed as LMIs.
- Once formulated in terms of LMIs, a problem can be solved exactly by efficient convex optimization algorithms (the "LMI solvers").
- While most problems with multiple constraints or objectives lack analytical solutions in terms of matrix equations, they often remain tractable in the LMI framework. This makes LMI-based design a valuable alternative to classical "analytical" methods.

For systems and control, the importance of LMI optimization stems from the fact that a wide variety of system and control problems can be recast as LMI problems. Therefore recasting a control problem as an LMI problem is equivalent to finding a "solution" to the original problem.

A linear matrix inequality has the form [98]

$$F(x) = F_0 + \sum_{i=1}^{m} x_i F_i > 0, \tag{A.1}$$

where $x \in \Re^m$ is the variable to be determined and symmetric matrices $F_i = F_i^T \in \Re^{n\times n}$, $i = 0,\cdots,m$, are given. The inequality symbol in (A.1) means that $F(x)$ is positive definite, i.e., $u^T F(x)u > 0$ for all nonzero $u \in \Re^n$.

Even though this canonical expression (A.1) is generic, LMIs rarely arise in this form in control applications. Instead, structured representation of LMIs is often used. For instance, the expression $A^T P + PA < 0$ in the Lyapunov inequality is explicitly described as a function of the matrix variable P, and A is the given matrix. In addition to saving notation, the structured representation may lead to more efficient computation.

Here we list some common issues that are standard in LMI text. We will encounter them throughout the book.

A.1.1 LMI Problems

Given an LMI $F(x) > 0$, the LMI problem [98] is to find x^{feas} such that $F(x^{feas}) > 0$ or determine that the LMI is infeasible. This is a convex feasibility problem.

As an example, consider the simultaneous Lyapunov stability problem [100]. We are given $A_i \in \Re^{n\times n}$, $i = 1,\cdots,r$, and need to find P satisfying the LMI:

$$P > 0, \qquad A_i^T P + PA_i < 0,\ i = 1,\cdots,r \tag{A.2}$$

or determine that no such P exists.

The LMI problems are tractable from both theoretical and practical viewpoints. They can be solved in polynomial time, and they can be solved in practice very efficiently by means of some of the most powerful tools available to date in the mathematical programming literature (e.g., the recently developed interior-point methods [101]).

The stability and stabilizability conditions encountered in this book are expressed in the form of LMIs. This recasting is significant in the sense that efficient convex optimization algorithms can be used for the stability analysis and control design. The recasting therefore constitutes solutions to the stability analysis and control design problems in the framework of the T-S fuzzy model.

LMI Control Toolbox [99], which was developed by The MathWorks™, provides state-of-the-art tools for the LMI-based analysis and design of control systems. Moreover, it offers a flexible and user-friendly environment to specify and solve general LMI problems (the LMI Lab). We choose it as our basic tool to solve LMI problems in our research.

A.1.2 The Schur Complement

The Schur complement [98] is standard in LMI context. The basic idea is as follows: the LMI

$$\begin{bmatrix} Q(x) & S(x) \\ S(x)^T & R(x) \end{bmatrix} > 0, \tag{A.3}$$

where $Q(x) = Q(x)^T$, $R(x) = R(x)^T$, and $S(x)$ depend affinely on x, is equivalent to

$$R(x) > 0, \qquad Q(x) - S(x)R(x)^{-1}S(x)^T > 0. \tag{A.4}$$

In other words, the set of nonlinear inequalities (A.4) can be represented as the LMI (A.3).

A.2 Continuous-time Markov Process

A Markov process [109, 110, 107] is, as translations from Russian state, "a process without after-effect". This means that the process has no memory, save the memory of the last observed point. In probability theory, a continuous-time Markov process is a stochastic process $\{X(t) : t \geq 0\}$ that satisfies the Markov property and takes values from a set called the state space. The Markov property states that at any times $s > t > 0$, the conditional probability distribution of the process at time s given the whole history of the process up to and including time t, depends only on the state of the process at time t. Mathematically, if $\{X(t), t > 0\}$, is a stochastic process, the Markov property states that

$$Pr[X(t+h) = y \mid X(s) = x(s), \forall s \leq t] = Pr[X(t+h) = y \mid X(t) = x(t)], \forall h > 0. \tag{A.5}$$

It is not hard to see from (A.5) that in effect, the state of the process at time s is conditionally independent of the history of the process before time t, given the state of the process at time t. Intuitively, one can define a time-homogeneous Markov process as follows.

Let $X(t)$ be the random variable describing the state of the process at time t. Now prescribe that in some small increment of time from t to $t+h$, the probability that the process makes a transition to some state j, given that it started in some state $i \neq j$ at time t, is given by

$$Pr(X(t+h) = j \mid X(t) = i) = q_{ij}h + o(h), \tag{A.6}$$

where $o(h)$ represents a quantity that goes to zero faster than h as h goes to zero. Hence, over a sufficiently small interval of time, the probability of a particular transition is roughly proportional to the duration of that interval.

Continuous-time Markov processes are most easily defined by specifying the transition rates q_{ij}, and these are typically given as the ij-th elements of the transition rate matrix Q.

In our research , we consider the most intuitive continuous-time Markov processes with Q-matrices that are:

- conservative - the i-th diagonal element q_{ii} of Q is given by $q_{ii} = -\sum_{i \neq i} q_{ij}$,
- stable - for any given state i, all elements q_{ij} (and q_{ii}) are finite.

Dynkin's formula

Let $X(t)$ be a continuous Markov process [108] and let $\mathbf{E}\{\cdot\}$ denote expectation operation. Also, let $\mathscr{A}$ be the infinitesimal generator of $X(t)$, defined by its action on compactly-supported C^2 (twice differentiable with continuous second derivative) functions $f : \mathbf{R}^n \to \mathbf{R}$ as

$$\mathscr{A} f(X) = \lim_{\tau \to 0} \frac{\mathbf{E}\{f(X(t+\tau))|X(t)\} - f(X(t))}{\tau}.$$

Let τ be a random time with $\mathbf{E}\{\tau\} < \infty$, then:

$$\mathbf{E}\{f(X(\tau))|X_0\} - f(X_0) = \mathbf{E}\{\int_0^\tau \mathscr{A} f(X_s) ds\}. \tag{A.7}$$

A.3 Lemmas

The following lemmas will be used in this research work.

Lemma A.1. *[111] For constant matrices H and E, a symmetric matrix G, and scalar $\varepsilon > 0$, the following inequality holds:*

$$G + HFE + E^T F^T H^H < 0,$$

where F satisfies $F^T F \leq I$, if and only if for any $\varepsilon > 0$,

$$G + \varepsilon H H^T + \varepsilon^{-1} E^T E < 0.$$

Lemma A.2. *[112] (Moon's inequality) For vectors* $\mathbf{a}$, $\mathbf{b} \in \Re^n$, *symmetric matrix $P \in \Re^{n \times n} > 0$ and scalar $\varepsilon > 0$, we have:*

$$-2\mathbf{a}^T \mathbf{b} \leq \frac{1}{\varepsilon} \mathbf{a}^T P^{-1} \mathbf{a} + \varepsilon \mathbf{b}^T P \mathbf{b}.$$

Lemma A.3. *[113] (Gronwall–Bellaman inequalities) Let $u(t)$ and $b(t)$ be nonnegative continuous functions for $t \geq \alpha$, and let*

$$u(t) \leq a + \int_\alpha^t b(s) u(s) ds, \ t \geq \alpha,$$

where $a \geq 0$ is a constant, then

$$u(t) \leq a \exp(\int_\alpha^t b(s) ds), \ t \geq \alpha.$$

References

1. Chan, H., Üzgüner, U.: Closed-loop control of systems over a communication network with queues. International Journal of Control 62, 493–510 (1995)
2. Etkin, B., Reid, L.D.: Dynamics of Flight: Stability and Control. Wiley, New York (1996)
3. Lee, K.B., Schneeman, R.D.: Internet-based distributed measurement and control applications. IEEE Instrumentation & Measurement Magazine 2, 23–27 (1999)
4. Baruch, J.E.F., Cox, M.J.: Remote control and robots: an Internet solution. Computing & Control Engineering Journal 7, 39–45 (1996)
5. Chow, M.Y., Tipsuwan, T.: Network-based control systems: A tutorial. In: Proceedings of 27th IEEE Annual Conference of the IEEE Industrial Electronics Society, pp. 1593–1602 (2001)
6. Zhang, W., Branicky, M.S., Phillips, S.M.: Stability of networked control systems. IEEE Control System Magzine 21, 84–99 (2001)
7. Hristu-Varsakelis, D., Levine, W.S. (eds.): Handbook of Networked and Embedded Control Systems. Birkhaüser, Boston (2005)
8. Halevi, Y., Ray, A.: Integrated communication and control systems: Part I - Analysis. Journal of Dynamic Systems, Measurement and Control 110, 367–373 (1988)
9. Ye, H., Walsh, G., Bushnell, L.: Wireless local area networks in the manufacturing industry. In: Proceedings of American Control Conference, Chicago, IL, pp. 2363–2367 (2000)
10. Ray, A.: Distributed data communication networks for real-time process control. Chem. Eng. Commun. 65, 139–154 (1988)
11. Leung, W.-L.D., Vanijjirattikhan, R., Li, Z., Xu, L., Richards, T., Ayhan, B., Chow, M.Y.: Intelligent space with time sensitive applications. In: Proceedings of IEEE/ASME International Conference on Advanced Intelligent Mechatronics, Monterey, California, USA, pp. 1413–1418 (2005)
12. Üzgüner, U., Göktas, H., Chan, H., Winkelman, J., Liubakka, M., Krotolica, R.: Automotive suspension control through a computer communication network. In: Proceedings of IEEE Conference on Control Applications, pp. 895–900 (1992)
13. Kondraske, G.V., Volz, R.A., Johnson, D.H., Tesar, D., Trinkle, J.C., Price, C.R.: Network-based infrastructure for distributed remote operations and robotics research. IEEE Transactions on Robotics and Automation 9, 702–704 (1993)
14. Evans, J., Filsfils, C.: Deploying IP and MPLS QoS for Multiservice Networks: Theory and Practice. Morgan Kaufmann, San Francisco (2007)

15. Tanenbaum, A.S.: Computer Networks, 4th edn. Prentice Hall, Upper Saddle River (2003)
16. Tangemann, M., Sauer, K.: Performance analysis of the timed token protocol of FDDI and FDDI-II. IEEE Journal on Selected Areas in Communications 9, 271–278 (1991)
17. Anderson, D.: FireWire System Architecture. Addison-Wesley, Reading (1998)
18. Hong, S.H.: Scheduling algorithm of data sampling times in the integrated communication and control systems. IEEE Transactions on Control Systems Technology 3, 225–230 (1995)
19. Ray, A., Halevi, Y.: Integrated communication and control systems: Part II - Design considerations. Journal of Dynamic Systems, Measurement and Control 110, 374–381 (1988)
20. Nilsson, J., Bernhardsson, B., Wittenmark, B.: Stochastic analysis and control of real-time systems with random time delays. Automatica 34, 57–64 (1998)
21. Nilsson, J.: Real-time Control Systems with Delays. Ph.D. Thesis, Department of Automatic Control, Lund Institute of Technology (1998)
22. Lawrenz, W.: CAN System Engineering: From Theory to Practical Applications. Springer, New York (1997)
23. Bertsekas, D., Gallager, R.: Data Networks, 2nd edn. Prentice Hall, Englewood Cliffs (1992)
24. Yu, M., Wang, L., Chu, T.: Sampled-data stabilisaztion of networked control systems with nonlinearity. IEE Proceedings of Control Theory Applications 162, 609–614 (2005)
25. Branicky, M.S., Phillips, S.M., Zhang, W.: Stability of networked control systems: Explicit analysis of delay. In: Proceedings of the America Control Conference, Chicago, IL, USA, pp. 2352–2357 (June 2000)
26. Naghshtabrizi, P., Hespanha, J.P.: Designing an observer-based controller for a network controk system. In: Proceedings of the 44th IEEE Conference on Decision and Control, Seville, Spain, pp. 848–853 (December 2005)
27. Montestruque, L.A., Antsaklis, P.: Stability of model-based networked control systems with time varying transmission time. IEEE Transactions on Automatic Control 49, 1562–1572 (2004)
28. Gao, H., Chen, T., Lam, J.: A new delay system approach to network-based control. Automatica 44, 39–52 (2008)
29. Montestruque, L.A., Antsaklis, P.: On the model-based control of networked systems. Automatica 39, 1837–1843 (2003)
30. Yue, D., Han, Q.-L.: Network-based robust $\mathscr{H}_\infty$ filtering for uncertain linear systems. IEEE Transactions on Signal Processing 54, 4293–4301 (2006)
31. Petersen, I.R., Savkin, A.V.: Multi-rate stabilization of multivariable discrete-time linear systems via a limited capacity communication channel. In: Proceedings of the 40th IEEE Conference on Decision and Control, Orlando, FL, USA, pp. 304–309 (December 2001)
32. Sahebsara, M., Chen, T., Shah, S.: Optimal $\mathscr{H}_2$ filtering in networked control systems with multiple packet dropout. IEEE Transactions on Automatic Control 52, 1508–1513 (2007)
33. Huo, Z., Fang, H.: Robust $\mathscr{H}_\infty$ filter design for networked control system with random time delays. In: Proceedings of 10th IEEE International Engineering of Complex Computer Systems, pp. 333–340 (2005)

34. Lu, L., Xie, L., Fu, M.: Optimal control of networked systems with limited communication: a combined heuristic and convex optimization approach. In: Proceedings of 42nd IEEE Conference on Decision and Control, Maui, Hawaii, USA, pp. 1194–1199 (December 2003)
35. Gao, H., Chen, T.: $\mathscr{H}_\infty$ estimation for uncertain systems with limited communication capacity. IEEE Transactions on Automatic Control 52, 2070–2084 (2007)
36. Wong, W.S., Brockett, R.W.: Systems with finite communication bandwidth constraints - Part II, Stabilization with limited information feedback. IEEE Transactions on Automatic Control 44, 1049–1053 (1999)
37. Ray, A., Liou, L.W., Shen, J.H.: State estimation using randomly delayed measurements. ASME Journal of Dynamic Systems, Measurement and Control 115, 19–26 (1993)
38. Wong, W.S., Brockett, R.W.: Systems with finite communication bandwidth constraints - Part I, State estimation problems. IEEE Transactions on Automatic Control 42, 1294–1299 (1997)
39. Elia, N., Mittler, S.: Stabilization of linear systems with limited information. IEEE Transactions on Automatic Control, 1384–1400 (2001)
40. Hu, L.-S., Bai, T., Shi, P., Wu, Z.: Sampled-data control of networked linear control systems. Automatica 43, 903–911 (2007)
41. Walsh, G.C., Ye, H., Bushnell, L.G.: Stability analysis of networked control systems. IEEE Transactions on Control Systems Technology 10, 438–446 (2002)
42. Fridman, E., Seuret, A., Richard, J.-P.: Robust sampled-data stabilization of linear systems: An input delay approach. Automatica 40, 1441–1446 (2004)
43. Åström, K., Wittenmark, B.: Adaptive Control. Addison-Wesley, Reading (1989)
44. Lian, F.-L., Moyne, J., Tilbury, D.: Modelling and optimal controller design of networked control systems with multile delays. International Journal of Control 76, 591–606 (2003)
45. Zhang, L., Hristu-Varsakelis, D.: Communication and control co-design for networked control systems. Automatica 42, 953–958 (2006)
46. Lin, H., Zhai, G., Antsaklis, P.J.: Robust stability and disturbance attenuation analysis of a class of networked control systems. In: Proceedings of 42nd IEEE Conference on Decision and Control, Maui, Hawaii, USA, pp. 1182–1187 (December 2003)
47. Decarlo, R.A., Branicky, M.S., Pettersson, S., Lennartson, B.: Perspectives and results on the stability and stabilizability of hybrid systems. Proceedings of the IEEE 88, 1069–1082 (2000)
48. Liberzon, D., Morse, A.S.: Basic problems in stability and design of switched systems. IEEE Control Systems Magazine 19, 59–70 (1999)
49. Branicky, M.S.: Multiple Lyapunov functions and other analysis tools for switched and hybrid systems. IEEE Transactions on Automatic Control 43, 475–482 (1998)
50. Kim, Y.H., Park, H.S., Kwon, W.H.: Stability and a scheduling method for network-based control systems. In: Proceedings of the 1996 IEEE IECON 22nd International Conference on Industrial Electronics, Control, and Instrumentation, pp. 934–939 (August 1996)
51. Park, H.S., Kim, Y.H., Kim, D.-S., Kwon, W.H.: A scheduling method for network-based control systems. IEEE Transactions on Control Systems Technology 10, 318–330 (2002)
52. Liou, L.-W., Ray, A.: Integrated communication and control systems: Part III - Non-identical sensor and controller sampling. Journal of Dynamic Systems, Measurement and Control 112, 357–364 (1990)

53. Nilsson, J., Bernhardsson, B.: LQR control over a Markov Communication Network. In: Proceedings of the 36th IEEE Conference on Decision and Control, San Diego, California, USA, Atlantis, pp. 4586–4591 (December 1997)
54. Walsh, G.C., Beldiman, O., Bushnell, L.: Asymptotic behavior of networked control systems. In: Proceedings of the 1999 IEEE International Conference on Control Applications, pp. 1448–1453 (August 1999)
55. Walsh, G.C., Ye, H., Bushnell, L.: Stability analysis of networked control systems. In: Proceedings of the America Control Conference, San Diego, CA, USA, pp. 2876–2880 (June 1999)
56. Walsh, G.C., Beldiman, O., Bushnell, L.: Error encoding algorithms for networked control systems. In: Proceedings of the 38th IEEE Conference on Decision and Control, Phoenix, Arizona, USA, pp. 4933–4938 (December 1999)
57. Luck, R., Ray, A.: An observer-based compensator for distributed delays. Automatica 26, 903–908 (1990)
58. Luck, R., Ray, A.: Experimental verification of a delay compensation algorithm for integrated communication and control systems. International Journal of Control 59, 1357–1372 (1994)
59. Goodwin, G.C., Haimovich, H., Quevedo, D.E., Welsh, J.S.: A moving horizon approach to networked control system design. IEEE Transactions on Automatic Control 49, 1427–1445 (2004)
60. Ferrari-Trecate, G., Mignone, D., Morari, M.: Moving horizon estimation for hybrid systems. IEEE Transactions on Automatic Control 47, 1663–1676 (2002)
61. Maciejowski, J.M.: Predictive Control with Constraints. Prentice-Hall, Upper Saddle River (2002)
62. Takagi, T., Sugeno, M.: Fuzzy identification of systems and its applications to medeling and control. IEEE Transactions on Systems, Man, and Cybernetics-Part B: Cybernetics 15, 116–132 (1985)
63. Zheng, Y., Fang, H., Wang, H.O.: Takagi-Sugeno fuzzy-model-based fault detection for networked control systems with Markov delays. IEEE Transactions on Systems, Man, and Cybernetics-Part B: Cybernetics 36, 924–929 (2006)
64. Wu, F., Sun, F., Liu, H., Sun, Z.: Guaranteed cost control for NCSs via a discrete-time jump fuzzy system approach. In: Proceedings of 2005 IEEE Conference on Networking, Sensing and Control, pp. 502–507 (2005)
65. Lee, K.C., Lee, S., Lee, M.H.: Remote fuzzy logic control of networked control system via Profibus-DP. IEEE Transactions on Industrial Electronics 50, 784–792 (2003)
66. Krasovskii, N.N., Lidskii, E.A.: Analytical design of controllers in systems with random attributes I, II, III. Automation Remote Control 22, 1021–1-025, 1141–1146, 1289–1294 (1961)
67. Srichander, R., Walker, B.K.: Stochastic stability analysis for continuous-time fault tolerant control systems. International Journal of Control 57(2), 433–452 (1993)
68. Nguang, S.K., Assawinchaichote, W., Shi, P., Shi, Y.: Robust $\mathscr{H}_\infty$ control design for uncertain fuzzy systems with Markovian jumps: an LMI approach. In: Proceedings of the American Control Conference, Portland, OR, USA, pp. 1805–1810 (June 2005)
69. Assawinchaichote, W., Nguang, S.K., Shi, P., Mizumoto, M.: Robust $\mathscr{H}_\infty$ control design for fuzzy singularly perturbed systems with Markovian jumps: an LMI approach. In: Proceedings of the 43rd IEEE Conference on Decision and Control, Atlantis, Paradise Island, Bahamas, pp. 803–808 (December 2003)
70. Ji, Y., Chizeck, H.J.: Controllability, stabilizability, and continuous-time Markovian jump linear quadratic control. IEEE Transactions on Automatic Control 35(7), 777–788 (1990)

71. Xu, S., Chen, T., Lam, J.: Robust $\mathscr{H}_\infty$ filtering for uncertain Markovian jump systems with mode-dependent time delays. IEEE Transactions on Automatic Control 48, 900–907 (2003)
72. Boukas, E.K., Liu, Z.K.: Robust stability and stabilizability of Markov jump linear uncertain systems with mode-dependent time delays. Journal of Optimal Theory and Application 109, 587–600 (2001)
73. Cao, Y.Y., Lam, J.: Robust $\mathscr{H}_\infty$ control of uncertain Markovian jump systems with time-delay. IEEE Transactions on Automatic Control 45, 77–83 (2000)
74. Zhang, L., Shi, Y., Chen, T., Huang, B.: A new method for stabilisation of networked control systems with random delays. IEEE Transactions on Automatic Control 50, 1177–1181 (2005)
75. Ji, Y., Chizeck, H.J.: Jump linear quadratic Gaussian control in continuous time. IEEE Transactions on Automatic Control 37, 1884–1892 (1992)
76. Mariton, M., Bertrand, P.: Output feedback control for a class of linear systems with jump parameters. IEEE Transactions on Automatic Control 30, 898–900 (1985)
77. Boukas, E.K., Liu, Z.K., Shi, P.: Delay-dependant stability and output feedback stabilisation of Markov jump systems with time-delay. IEE Proc. Control Theory Applications 149, 379–386 (2002)
78. Chen, W.-H., Xu, J.-X., Guan, Z.-H.: Guaranteed cost control foe uncertain Markovian jump systems with mode-dependent time-delays. IEEE Transactions on Automatic Control 48, 2270–2277 (2003)
79. de Souza, C.E., Trofino, A., Barbosa, K.A.: Mode-independent $\mathscr{H}_\infty$ filters for hybrid Markov linear systems. In: Proceedings of 43rd IEEE Conference on Decision and Control, Atlantis, Paradise Island, Bahamas, pp. 947–952 (December 2004)
80. Jeung, E.T., Kim, J.H., Park, H.B.: $\mathscr{H}_\infty$-output feedback controller design for linear systems with time-varying delayed state. IEEE Transactions on Automatic Control 43, 971–974 (1998)
81. He, J.B., Wang, Q.G., Lee, T.H.: $\mathscr{H}_\infty$ disturbance attenuation for state delayed systems. System Control Letters 33, 105–114 (1998)
82. Zames, G.: Feedback and optimal sensitivity: model reference transformations, multiplicative seminorms, and approximate inverse. IEEE Transactions on Automatic Control 26, 301–320 (1981)
83. Stoorvogel, A.: The $\mathscr{H}_\infty$ Control Problem: A State Space Approach. Prentice Hall, Englewood Cliffs (1992)
84. Choi, H.H., Chung, M.J.: Memoryless $\mathscr{H}_\infty$ controller design for linear systems with delayed state and control. Automatica 31, 917–919 (1995)
85. Jeung, E.T., Kim, J.H., Park, H.B.: $\mathscr{H}_\infty$ output feedback controller design for linear systems with time-varying delayed state. IEEE Transactions on Automatic Control 43, 971–974 (1998)
86. Anderson, B., Vongpantlerd, S.: Network Analysis and Synthesis: A Modern System Theory Appraoch. Prentice-Hall, Inc., New Jersey (1973)
87. Kokame, H., Kobayashi, H., Mori, T.: Robust $\mathscr{H}_\infty$ performance for linear delay-differential systems with time-varying uncertainties. IEEE Transactions on Automatic Control 43, 223–226 (1998)
88. Basseville, M.: Detecting changes in signals and systems-a survey. Automatica 24, 309–326 (1998)
89. Isermann, R.: Process fault detection based on modelling and estimation methods-a survey. Automatica 20, 387–404 (1984)
90. Frank, P.M.: Fault diagnosis in dynamic systems using analytical and knowledge based redundancy-a survey of some new results. Automatica 26, 459–474 (1990)

91. Nguang, S.K., Shi, P., Ding, S.: Delay-dependent fault estimation for uncertain time-delay nonlinear systems: An LMI approach. International Journal of Robust and Nonlinear Control 16, 913–933 (2006)
92. Wang, H., Wang, C., Gao, H., Wu, L.: An LMI approach to fault detection and isolation filter design for Markovian jump system with mode-dependent time-delays. In: Proceedings of the 2006 American Control Conference, Minneapolis, Minnesota, USA, pp. 5686–5691 (June 2006)
93. Ding, S.X., Zhong, M., Tang, B., Zhang, P.: An LMI approach to the design of fault detection filter for time-delay LTI systems with unknown inputs. In: Proceedings of the 2001 American Control Conference, Arlington, VA, USA, pp. 1467–1472 (December 2003)
94. Zhong, M., Ye, H., Shi, P., Wang, G.: Fault detection for Markovian jump systems. IEE Proceedings of Control Theory Applications 152, 397–402 (2005)
95. Blanke, M., Kinnaert, M., Lunze, J., Staroswiecki, M.: Diagnosis and Fault-Tolerant Control. Springer, Heidelberg (2006)
96. Hale, J.K., Lunel, S.M.V.: Introduction to Functional Differential Equations. Springer, New York (1993)
97. Teel, A.R.: Connections between Razumikhin-type theorems and the ISS nonlinear small gain theorem. IEEE Transactions on Automatic Control 43, 960–964 (1998)
98. Boyd, S., Ghaoui, L., Feron, E., Balakrishnan, V.: Linear Matrix Inequalities in System and Control Theory. SIAM, Philadelphia (1994)
99. Gahinet, P., Nemirovski, A., Laub, A.J., Chilali, M.: LMI Control Toolbox. The MathWorks, Inc., Natick (1995)
100. Tanaka, K., Sugeno, M.: Stability analysis of fuzzy systems using Lyapunov's direct method. In: Proc. NAFIPS 1990, pp. 133–136 (1990)
101. Nesterov, Y., Nemirovsky, A.: Interior-Point Polynomial Methods in Convex Programming. SIAM, Philadelphia (1994)
102. Mikheev, Y.V., Sobolev, V.A., Fridman, E.M.: Asymptotic analysis of digital control systems. Automation and Remote Control 49, 1175–1180 (1980)
103. Fridman, E.: Use of models with aftereffect in the problem of design of optimal digital control. Automation and Remote Control 53, 1523–1528 (1992)
104. Hale, J.: Theory of Functional Differential Equations. Springer, New York (1977)
105. Mao, X.: Stochastic funtional differential equations with Markovian switching. Functional Differential Equations 6, 375–396 (1999)
106. Niculescu, S.I., Fu, M., Li, H.: Delay-dependent closed-loop stability of linear systems with input delays: An LMI approach. In: Proceedings of the 36th IEEE Conference on Decision and Control, San Diego, CA, USA, pp. 1623–1628 (December 1997)
107. Dynkin, E.B.: Markov processes. Academic Press Inc., New York (1965)
108. Øksendal, B.K.: Stochastic differential equations: an introduction with applications. Springer, Berlin (2003)
109. Doob, J.L.: Stochastic Processes. Wiley, New York (1956)
110. Kushner, H.J.: Stochastic Stability and Control. Academic Press, New York (1967)
111. Wang, Y., Xie, L., De Souza, C.E.: Robust control of a class of uncertain nonlinear systems. System Control Letters 19, 139–149 (1992)
112. Moon, Y.S., Park, P., Kwon, W.H., Lee, Y.S.: Delay-dependent robust stabilization of uncertain state-delayed systems. International Journal of Control 74, 1447–1455 (2001)
113. Bainov, D., Simeonov, P.: Integral Inequalities and Applications. Kluwer Academic Publishers, Dordrecht (1992)
114. Anderson, B., Vongpantlerd, S.: Network Analysis and Synthesis: A Modern System Theory Approach. Prentice-Hall, Inc., New Jersey (1973)

115. Nagpal, K.M., Khargonekar, P.P.: Filtering and smoothing in an $\mathscr{H}_\infty$ setting. IEEE Transactions on Automatic Control 36, 152–166 (1991)
116. Chen, J., Patton, R.J.: Robust Model-based Fault Diagnosis for Dynamic Systems. Kluwer Academic Publishers, Boston (1999)
117. Gertler, J.: Fault Detection and Diagnosis in Engineering. Marcel Dekker, New York (1998)
118. Chen, J., Patton, R.J.: Standard $\mathscr{H}_\infty$ filtering formulation of robust fault detection. In: Proc. Safe Process, Budapest, Hungary, pp. 256–261 (2000)
119. Ding, S.X., Jeinsch, T., Frank, P.M., Ding, E.L.: A unified approach to the optimization of fault detection systems. Int. J. Adapt. Contr. Sig. Proc. 14, 725–745 (2000)
120. Patton, R.J., Hou, M.: On sensitivity of robust fault detection observers. In: Proc. 14th IFAC Congress, Beijing, China, pp. 67–72 (1999)
121. Zhong, M., Ding, S.X., Lam, J., Wang, H.: LMI approach to design robust fault detection filter for uncertain LTI systems. Automatica 39, 543–550 (2003)
122. Huang, D., Nguang, S.K.: Static output feedback controller design for fuzzy systems: An ILMI approach. Information Science 177, 3005–3015 (2007)
123. Cao, Y.Y., Lam, J., Sun, Y.X.: Static output feedback stabiliztion: an ILMI approach. Automatica 34, 1641–1645 (1998)
124. Taniguchi, T., Tanaka, K., Ohtake, H., Wang, H.: Model construction, rule reduction, and robust compensation for generalized form of takagi-sugeno fuzzy systems. IEEE Trans. Fuzzy Syst. 9, 525–538 (2001)
125. Assawinchaichote, W., Nguang, S.K., Shi, P.: Fuzzy Control and Filter Design for Uncertain Fuzzy Systems. Springer, Heidelberg (2006)
126. Lee, H.J., Park, J.B., Joo, Y.H.: Robust control for uncertain Takagi-Sugeno fuzzy systems with time-varying input delay. Journal of Dynamic Systems, Measurement, and Control 127, 302–306 (2005)

Index

$\mathscr{L}_2$ gain, 53, 65

balancing an inverted pendulum, 123

data packet dropouts, 3, 5, 19
disturbance attenuation, 7, 53
dynamic output feedback controller, 37
Dynkin's formula, 150

fault detection and isolation, 73
fault estimation error, 75
filter design, 65
filter error, 67
fuzzy disturbance attenuation, 117
fuzzy dynamic output feedback controller, 107
fuzzy fault estimator, 137
fuzzy filter design, 129
fuzzy state feedback controller, 93

Gronwall–Bellaman Inequalities, 150

if-then rules, 87
input delays, 53

linear matrix inequality, 12
linear quadratic gaussian, 9
low-frequency faults, 80
Lyapunov function, 8

Markov process, 11, 19, 25, 149
Markovian jump linear system, 25
mass-spring-damper mechanical system, 89, 102
medium access control, 5
membership function, 89
message rejection, 5
modeling of NCSs, 17
Moon's inequality, 150

network-induced delay, 3, 18
networked control systems, 1

Quality-of-Service, 2

random access network, 6
Razumikhin-type method, 11
Razumikhin-type theorem, 45, 56
robust fault estimator, 73

sampled-data systems, 7
sampling period, 4
scheduling network, 5
Schur complement, 45, 148
sector nonlinearity, 88
stability analysis, 7
state feedback controller, 25
stochastic process, 19
switched system, 8

T-S fuzzy model, 10, 87
time-delay systems, 7
transition probability matrix, 20

vacant sampling, 5

zero order hold, 4

Lecture Notes in Control and Information Sciences

Edited by M. Thoma, F. Allgöwer, M. Morari

Further volumes of this series can be found on our homepage:
springer.com

Vol. 386: Huang, D.;
Nguang, S.K.:
Robust Control for Uncertain Networked Control Systems with Random Delays
159 p. 2009 [978-1-84882-677-9]

Vol. 385: Jungers, R.:
The Joint Spectral Radius
144 p. 2009 [978-3-540-95979-3]

Vol. 384: Magni, L.; Raimondo, D.M.;
Allgöwer, F. (Eds.):
Nonlinear Model Predictive Control
572 p. 2009 [978-3-642-01093-4]

Vol. 383: Sobhani-Tehrani E.;
Khorasani K.;
Fault Diagnosis of Nonlinear Systems
Using a Hybrid Approach
360 p. 2009 [978-0-387-92906-4]

Vol. 382: Bartoszewicz A.;
Nowacka-Leverton A.;
Time-Varying Sliding Modes for Second and Third Order Systems
192 p. 2009 [978-3-540-92216-2]

Vol. 381: Hirsch M.J.; Commander C.W.;
Pardalos P.M.; Murphey R. (Eds.)
Optimization and Cooperative Control Strategies: Proceedings of the 8th International Conference on Cooperative Control and Optimization
459 p. 2009 [978-3-540-88062-2]

Vol. 380: Basin M.
New Trends in Optimal Filtering and Control for Polynomial and Time-Delay Systems
206 p. 2008 [978-3-540-70802-5]

Vol. 379: Mellodge P.; Kachroo P.;
Model Abstraction in Dynamical Systems:
Application to Mobile Robot Control
116 p. 2008 [978-3-540-70792-9]

Vol. 378: Femat R.; Solis-Perales G.;
Robust Synchronization of Chaotic Systems
Via Feedback
199 p. 2008 [978-3-540-69306-2]

Vol. 377: Patan K.
Artificial Neural Networks for
the Modelling and Fault
Diagnosis of Technical Processes
206 p. 2008 [978-3-540-79871-2]

Vol. 376: Hasegawa Y.
Approximate and Noisy Realization of
Discrete-Time Dynamical Systems
245 p. 2008 [978-3-540-79433-2]

Vol. 375: Bartolini G.; Fridman L.; Pisano A.;
Usai E. (Eds.)
Modern Sliding Mode Control Theory
465 p. 2008 [978-3-540-79015-0]

Vol. 374: Huang B.; Kadali R.
Dynamic Modeling, Predictive Control
and Performance Monitoring
240 p. 2008 [978-1-84800-232-6]

Vol. 373: Wang Q.-G.; Ye Z.; Cai W.-J.;
Hang C.-C.
PID Control for Multivariable Processes
264 p. 2008 [978-3-540-78481-4]

Vol. 372: Zhou J.; Wen C.
Adaptive Backstepping Control of Uncertain Systems
241 p. 2008 [978-3-540-77806-6]

Vol. 371: Blondel V.D.; Boyd S.P.;
Kimura H. (Eds.)
Recent Advances in Learning and Control
279 p. 2008 [978-1-84800-154-1]

Vol. 370: Lee S.; Suh I.H.;
Kim M.S. (Eds.)
Recent Progress in Robotics:
Viable Robotic Service to Human
410 p. 2008 [978-3-540-76728-2]

Vol. 369: Hirsch M.J.; Pardalos P.M.;
Murphey R.; Grundel D.
Advances in Cooperative Control and Optimization
423 p. 2007 [978-3-540-74354-5]

Vol. 368: Chee F.; Fernando T.
Closed-Loop Control of Blood Glucose
157 p. 2007 [978-3-540-74030-8]

Vol. 367: Turner M.C.; Bates D.G. (Eds.)
Mathematical Methods for Robust and Nonlinear Control
444 p. 2007 [978-1-84800-024-7]

Vol. 366: Bullo F.; Fujimoto K. (Eds.)
Lagrangian and Hamiltonian Methods for
Nonlinear Control 2006
398 p. 2007 [978-3-540-73889-3]

Vol. 365: Bates D.; Hagström M. (Eds.)
Nonlinear Analysis and Synthesis Techniques for Aircraft Control
360 p. 2007 [978-3-540-73718-6]

Vol. 364: Chiuso A.; Ferrante A.; Pinzoni S. (Eds.)
Modeling, Estimation and Control
356 p. 2007 [978-3-540-73569-4]

Vol. 363: Besançon G. (Ed.)
Nonlinear Observers and Applications
224 p. 2007 [978-3-540-73502-1]

Vol. 362: Tarn T.-J.; Chen S.-B.; Zhou C. (Eds.)
Robotic Welding, Intelligence and Automation
562 p. 2007 [978-3-540-73373-7]

Vol. 361: Méndez-Acosta H.O.; Femat R.; González-Álvarez V. (Eds.):
Selected Topics in Dynamics and Control of Chemical and Biological Processes
320 p. 2007 [978-3-540-73187-0]

Vol. 360: Kozlowski K. (Ed.)
Robot Motion and Control 2007
452 p. 2007 [978-1-84628-973-6]

Vol. 359: Christophersen F.J.
Optimal Control of Constrained Piecewise Affine Systems
190 p. 2007 [978-3-540-72700-2]

Vol. 358: Findeisen R.; Allgöwer F.; Biegler L.T. (Eds.): Assessment and Future Directions of Nonlinear Model Predictive Control
642 p. 2007 [978-3-540-72698-2]

Vol. 357: Queinnec I.; Tarbouriech S.; Garcia G.; Niculescu S.-I. (Eds.):
Biology and Control Theory: Current Challenges
589 p. 2007 [978-3-540-71987-8]

Vol. 356: Karatkevich A.:
Dynamic Analysis of Petri Net-Based Discrete Systems
166 p. 2007 [978-3-540-71464-4]

Vol. 355: Zhang H.; Xie L.:
Control and Estimation of Systems with Input/Output Delays
213 p. 2007 [978-3-540-71118-6]

Vol. 354: Witczak M.:
Modelling and Estimation Strategies for Fault Diagnosis of Non-Linear Systems
215 p. 2007 [978-3-540-71114-8]

Vol. 353: Bonivento C.; Isidori A.; Marconi L.; Rossi C. (Eds.)
Advances in Control Theory and Applications
305 p. 2007 [978-3-540-70700-4]

Vol. 352: Chiasson, J.; Loiseau, J.J. (Eds.)
Applications of Time Delay Systems
358 p. 2007 [978-3-540-49555-0]

Vol. 351: Lin, C.; Wang, Q.-G.; Lee, T.H., He, Y.
LMI Approach to Analysis and Control of Takagi-Sugeno Fuzzy Systems with Time Delay
204 p. 2007 [978-3-540-49552-9]

Vol. 350: Bandyopadhyay, B.; Manjunath, T.C.; Umapathy, M.
Modeling, Control and Implementation of Smart Structures 250 p. 2007 [978-3-540-48393-9]

Vol. 349: Rogers, E.T.A.; Galkowski, K.; Owens, D.H.
Control Systems Theory and Applications for Linear Repetitive Processes
482 p. 2007 [978-3-540-42663-9]

Vol. 347: Assawinchaichote, W.; Nguang, K.S.; Shi P.
Fuzzy Control and Filter Design for Uncertain Fuzzy Systems
188 p. 2006 [978-3-540-37011-6]

Vol. 346: Tarbouriech, S.; Garcia, G.; Glattfelder, A.H. (Eds.)
Advanced Strategies in Control Systems with Input and Output Constraints
480 p. 2006 [978-3-540-37009-3]

Vol. 345: Huang, D.-S.; Li, K.; Irwin, G.W. (Eds.)
Intelligent Computing in Signal Processing and Pattern Recognition
1179 p. 2006 [978-3-540-37257-8]

Vol. 344: Huang, D.-S.; Li, K.; Irwin, G.W. (Eds.)
Intelligent Control and Automation
1121 p. 2006 [978-3-540-37255-4]

Vol. 341: Commault, C.; Marchand, N. (Eds.)
Positive Systems
448 p. 2006 [978-3-540-34771-2]

Vol. 340: Diehl, M.; Mombaur, K. (Eds.)
Fast Motions in Biomechanics and Robotics
500 p. 2006 [978-3-540-36118-3]

Vol. 339: Alamir, M.
Stabilization of Nonlinear Systems Using Receding-horizon Control Schemes
325 p. 2006 [978-1-84628-470-0]

Vol. 338: Tokarzewski, J.
Finite Zeros in Discrete Time Control Systems
325 p. 2006 [978-3-540-33464-4]

Vol. 337: Blom, H.; Lygeros, J. (Eds.)
Stochastic Hybrid Systems
395 p. 2006 [978-3-540-33466-8]

Vol. 336: Pettersen, K.Y.; Gravdahl, J.T.; Nijmeijer, H. (Eds.)
Group Coordination and Cooperative Control
310 p. 2006 [978-3-540-33468-2]